Disaster Risk Reduction

Methods, Approaches and Practices

Series Editor

Rajib Shaw, Keio University, Shonan Fujisawa Campus, Fujisawa, Japan

Disaster risk reduction is a process that leads to the safety of communities and nations. After the 2005 World Conference on Disaster Reduction, held in Kobe, Japan, the Hyogo Framework for Action (HFA) was adopted as a framework for risk reduction. The academic research and higher education in disaster risk reduction has made, and continues to make, a gradual shift from pure basic research to applied, implementation-oriented research. More emphasis is being given to multi-stakeholder collaboration and multi-disciplinary research. Emerging university networks in Asia, Europe, Africa, and the Americas have urged process-oriented research in the disaster risk reduction field. With this in mind, this new series will promote the output of action research on disaster risk reduction, which will be useful for a wide range of stakeholders including academicians, professionals, practitioners, and students and researchers in related fields. The series will focus on emerging needs in the risk reduction field, starting from climate change adaptation, urban ecosystem, coastal risk reduction, education for sustainable development, community-based practices, risk communication, and human security, among other areas. Through academic review, this series will encourage young researchers and practitioners to analyze field practices and link them to theory and policies with logic, data, and evidence. In this way, the series will emphasize evidence-based risk reduction methods, approaches, and practices.

Eleonora Gioia

Climate Change Adaptation in the Adriatic Coastal Areas

From European Strategies to Local Actions

Springer

Eleonora Gioia
Department of Life and Environmental Sciences
Marche Polytechnic University
Ancona, Italy

ISSN 2196-4106 ISSN 2196-4114 (electronic)
Disaster Risk Reduction
ISBN 978-981-97-6654-3 ISBN 978-981-97-6655-0 (eBook)
https://doi.org/10.1007/978-981-97-6655-0

Disclaimer: Funding for this work was provided by the EU Italy-Croatia Cross-border Co-operation Programme, which funded the RESPONSe (Strategies to adapt to climate change in Adriatic regions) project (ID10046849). The book reflects the author's views; the Programme authorities are not liable for any use that may be made of the information contained therein.

This Springer imprint is published by the registered company Springer Nature Singapore Pte Ltd.
The registered company address is: 152 Beach Road, #21-01/04 Gateway East, Singapore 189721, Singapore

To my beloved children, Sara and Alessandro.
To my caring husband, Rino.
May this book remind us all that, together, we can conceive a world where hope and sustainability thrive.

Foreword

Change is a lifelong companion of Homo sapiens, the true constant of the human experience on Planet Earth. Since its appearance, our species had to endure and adjust to continuous environmental transformations, natural or anthropic. The technological developments enabled humans to better adjust to these changes (i.e., warmer clothing for cold climate, toolmaking or buildings) and at the same time to transform our surrounding better to serve our needs (i.e., the agricultural revolution). Such human–environment interaction has been the focus of geographical research for long time, and various theories have tried to conceive overarching explanatory models (i.e., environmental determinism and environmental probabilism), yet the profound essence of these relations still remains elusive. Indeed, the increasing power of Homo sapiens to transform its surrounding is at issue (i.e., Anthropocene) with a cadre of uncertain and dramatic consequences. The very survival of our specie on Planet Earth will depend on our ability to adjust to natural changes and govern those changes caused by our development models.

Current research aims at defining a sustainable balance between resource consumption and habitat preservation, control of and adaptation to the natural processes. These are the essential conundrums of geography and human ecology and require reasoning across observational scales and disciplinary boundaries. To tackle the most evident challenge of the current environmental crisis, namely climate change, research must contemplate spatial and temporal analysis and cross-disciplinary boundaries to find new ways of survival.

This monographic volume by Eleonora Gioia is an example of this needed research, connecting global and local processes and confronting the issue of adaptive capacity to environmental changes. The book discusses lessons drawn from the research project "Strategies to adapt to climate change in Adriatic regions" (RESPONSe) financed by the European Union to promote climate-smart governance approaches and promote sustainable living in coastal areas. The research started a review of the adaptation strategies to climate change planned or adopted by the regions participating in the project, i.e., three Regions of Northern, Central, and Southern Italy (Friuli-Venezia Giulia, Marche, Puglia) and three Counties of Northern, Central and Southern Croatia (Primorje-Gorski Kotar, Šibenik-Knin,

Dubrovnik-Neretva). The analysis started at the European scale, to progressively focus on national and regional level, to finally arrive at the sub-regional level (e.g., group of municipalities) on both sides of the Adriatic coasts.

Undeniably, climate change has increased extreme precipitation, inland and coastal flooding as well as water scarcity and heat stress, which besides posing serious limitations in terms of sustainable development also pose paramount challenges in terms of Disaster Risk Reduction. Among the various approaches discussed in the book are the so-called nature-based solutions for climate change adaptation.

Despite the lack of clear academic evidence regarding the actual value of citizen involvement in the governance of such processes, in the project RESPONSe participatory approaches and stakeholders' inclusion appeared pivotal to push for transformative adaptation processes. Community participation in planning brings traditional and community knowledge into policies for risk governance, thus fostering greater integration between adaptation to climate change and disaster risk management.

This volume highlights that climate change adaptation and Disaster Risk Reduction face similar challenges as they both have to work toward changing the way people perceive their rights and duties toward nature and how they recognize and manage risks.

Fausto Marincioni
Professor of Environmental
Geography, Department of Life
and Environmental Sciences
Marche Polytechnic University
Ancona, Italy

Acknowledgements

This research was carried out at the Department of Life and Environmental Sciences of the Polytechnic University of Marche in Ancona (Italy). I would like to thank Prof. Fausto Marincioni and the Disaster Risk Reduction Laboratory of the Polytechnic University of Marche for their unfailing trust and valuable support. In particular, I would like to thank the reference points of this project, Dr. Alessandra Colocci, Dr. Cristina Casareale and, above all, Dr. Noemi Marchetti, without whom I would not have been able to complete this work.

I would also like to express my sincere gratitude to the RESPONSe project's partner organizations and to all the qualified people belonging to them; their irreplaceable collaboration has allowed us to achieve these excellent results together: Informest (Lead Partner), the Friuli-Venezia Giulia Energy Management Agency (APE-FVG), the Veneto Regional Agency for Environmental Protection and Prevention (ARPAV), the Apulia Region, the Hrvoje Požar Energy Institute (EIHP), the Croatian Meteorological and Hydrological Service (DHMZ) and the Institute of Oceanography and Fisheries (IOF), together with the Polytechnic University of Marche.

I would also like to thank my family, especially my husband Rino and my beloved children, Sara and Alessandro, for their love and constant support over the years.

Finally, I would like to thank the Springer Nature team, especially Yosuke Nishida and Sivananth S. Siva Chandran for the excellent cooperation.

Polverigi, Italy Eleonora Gioia

About This Book

Adaptation to climate change is critical for effective Disaster Risk Reduction, as the frequency and intensity of disasters are increasing and being amplified by climate-related factors.

This book examines climate change adaptation in the planning processes of Adriatic coastal communities in Italy and Croatia. It focuses on the urgency of climate change, particularly in marine and coastal areas, and addresses the need to identify, prepare for, and reduce climate impacts.

Building on the results of the EU Interreg Italy-Croatia RESPONSe project, it aims to improve local climate knowledge, engage stakeholders, and support local strategy development in northern, central, and southern Adriatic. It details strategies and multilevel governance—European, national, and regional—in addressing climate change impacts. By examining observed and projected climate risks, it evaluates the effectiveness of such strategies for both adaptation and mitigation efforts. The book culminates in an open database of best adaptation practices, tailored to the local context, to bridge policy gaps and promote cooperation in the Adriatic Basin.

More than just a compilation, the book offers a methodological framework for addressing implementation challenges within existing adaptation policies. It analyzes the complex relationships between levels of governance and encourages collaboration among policymakers, scientists, and citizens.

Contents

Chapter 1
Reducing the Risks Posed by Climate Change Through Multi-scalar, Multi-stakeholder Governance

Life on Earth is profoundly affected by weather and climate: they are essential to human daily experience, health, food production, and well-being, but it is also important to remember that humans can affect the climate as well. The delicate balance of our climate system, intricately intertwined by the interacting components of the atmosphere, hydrosphere, cryosphere, land surface and biosphere, has been significantly disturbed by human activities, which act as an external force of change within this intricate equilibrium (IPCC 2001). While the climate system is inherently dynamic, human interventions have caused perturbations that disrupt this balance.

Article 1 of the United Nations Framework Convention on Climate Change (UNFCCC) defines climate change as "a change in climate which is attributed, directly or indirectly, to human activities that modify the composition of the global atmosphere and which is further attributed to natural climate variability observed over comparable time periods" (UN General Assembly 1994). Indeed, while natural phenomena such as volcanic eruptions, earthquakes, and meteorological variations contribute to these changes, the greatest proportion of contemporary climate change is due to human-induced environmental changes. These are attributable to all anthropic activities which, through emissions, spills, exploitation, and alteration of the environmental heritage, induce damage and changes - reversible or non-reversible - in the composition of the atmosphere, soil, and sea. The Industrial Revolution marked the genesis of anthropogenic perturbations to atmospheric composition, an era in which human actions catalyzed a trajectory of change with far-reaching consequences (IPCC 2001). In addition, the Intergovernmental Panel on Climate Change has found that global warming has accelerated significantly in the second half of the twentieth century, and that humans are responsible for most of it (IPCC 2018). This change has manifested itself primarily through the emission of greenhouse gases, which has altered the radiative forcing of the system and caused a discernible shift in climate patterns and extremes (IPCC 2018; Rogelj et al. 2018). However, the ramifications of climate change extend far beyond mere statistical shifts; they manifest as tangible hazards that threaten both human well-being and ecological stability.

E. Gioia, *Climate Change Adaptation in the Adriatic Coastal Areas*, Disaster Risk Reduction, https://doi.org/10.1007/978-981-97-6655-0_1

The European Union (EU), among others, is facing the consequences of climate-related hazards, ranging from temperature extremes to devastating floods and droughts. These events cause significant economic losses, disrupt ecosystems, and pose serious threats to human health and infrastructure (EEA 2023). Between 1980 and 2022, the combined impact of these events has shown a steady increase in economic losses, demonstrating a worrying trend of escalating vulnerability in the face of climate change (EEA 2023). The economic toll of climate-related extremes in the EU, quantified in billions of euros, is a stark reminder of the increasing severity and frequency of these events. While some countries have made progress in insuring against climate-related losses, a considerable proportion of these losses remain uninsured, exacerbating the financial burden of these disasters (EEA 2023). Certainly, climate-related hazards are projected to intensify and amplify their impacts across multiple sectors and landscapes, underscoring the urgency of a unified, inclusive approach to tackling the problem. Comprehensive strategies aimed at strengthening societal resilience and adaptive capacity are therefore essential to address the environmental, social, and economic challenges posed by climate change. The solutions require not only concerted international cooperation, but also a paradigm shift towards sustainable practices, resilience-building measures, and holistic strategies to safeguard the future of the planet and the well-being of future generations. It may be important to clarify that strategies refer to the programmatic documents implemented by plans, whereas measures are actions that address specific problems within specific sectors and needs (e.g., creating or repairing wells for drinking and cooling to cope with heat waves in cities).

Addressing climate change requires a multifaceted approach that includes both mitigation and adaptation strategies at different geographical scales (Nalau et al. 2015).

In particular, mitigation is defined as the set of actions aimed at preventing or reducing emissions of greenhouse gases (GHG) into the atmosphere in order to reduce the impacts of climate change (EEA 2023). Mitigation is achieved either by reducing the sources of these gases (e.g., increasing the share of renewable energy, establishing a cleaner mobility system) or by increasing the storage of these gases (e.g., increasing the size of forests). Mitigation is therefore a human intervention that reduces the sources of GHG emissions and/or enhances the sinks through energy efficiency and decarbonization policies, spanning various sectors. This approach dates back to the 1997 Kyoto Protocol, an international agreement to combat global warming by limiting and reducing emissions. In the following years, many other regulations dealing with climate change and the reduction of carbon dioxide (CO_2) and other greenhouse gas emissions were implemented, such as the European Climate Change Program (ECCP) (European Commission 2000), the EU Emissions Trading System (European Commission 2003), the Climate and Energy Policy Framework 2030 (European Commission 2014).

On the other hand, adaptation seeks to take appropriate action to prevent or limit the damage from climate change, or to take advantage of the opportunities it may bring (EEA 2023). Examples of adaptation include large-scale infrastructure changes (e.g., building defenses to protect against sea level rise) as well as behavioral changes (e.g., individuals reducing their food waste). In essence, adaptation is the process of adjusting to the current and future impacts of climate change through risk assessment, policy engagement, and structural and non-structural interventions. Adaptation only began to appear in policy strategies as evidence of climate impacts began to mount, such as the melting of Arctic sea ice and mountain glaciers, the thawing of permafrost, extreme heat waves, floods, and storm damage (Biesbroek et al. 2009). Adaptation, indeed, is increasingly important for regions around the world where substantial changes in climate could have an impact on populations and human activities (Mathison et al. 2013). At the global level, the first references on adaptation are collected in the White Paper on Adaptation to Climate Change (European Commission 2009), the Cancun Adaptation Framework adopted in 2010 under the United Nations Framework Convention on Climate Change (UNFCCC), the EU Strategy on Adaptation to Climate Change (European Commission 2013a), and the Green Paper on Disaster Prevention and Insurance (European Commission 2013b).

Certainly, an effective response to climate change requires a balanced approach that uses mitigation to reduce emissions and adaptation to cope with unavoidable impacts, requiring tailored strategies for different geographic contexts. But both strategies face challenges. Mitigation, in particular, must be implemented on a global scale, and the scale of the effort severely limits its effectiveness. Conversely, conventional adaptation options are often already defined by regulations at different scales (supranational, national, and regional), and local specificities may limit the usefulness of generic adaptation options (Smith and Pilifosova 2001; Füssel 2007; Bormann et al. 2012; Rayner 2023). Although the analysis of the impacts of climate change on different sectors and the viable solutions for mitigation and adaptation have long been the subject of study in the scientific literature, the analysis of the needs for effective adaptation options in specific contexts is still difficult to determine (Arnell 2010).

Tackling climate change requires a multifaceted approach involving different levels of governance, from local to international (Adger et al. 2005). The cross-sectoral nature of such challenges requires integrated policies and coordinated actions across governments and sectors (Gioia et al. 2020). Therefore, climate change advocates a collaborative approach called multilevel governance, which emphasizes inclusive decision-making, integrated solutions, policy coherence, adaptability, and global cooperation (Pangalos 2023). Inclusive and participatory decision-making ensures that diverse perspectives are considered and that decisions are made collectively, acknowledging the complex nature of climate change. In addition, integrated and holistic solutions take into account the interconnectedness and interdependencies of the various systems involved in climate change, such as the ecological, social, economic, and political systems. In addition, policy coherence ensures the alignment and consistency of policies at multiple scales. This avoids policy inconsistencies and facilitates effective action on climate change. Multilevel governance also promotes adaptive, iterative, and flexible approaches to address the evolving circumstances, uncertainties, and risks associated with climate change. Finally, global cooperation plays a critical role. The United Nations Framework Convention on Climate Change (UNFCCC) and the Paris Agreement exemplify global efforts to address climate change by setting common goals and providing platforms for countries to negotiate and agree on actions. Indeed, climate change transcends national boundaries and requires a collaborative framework for communication, data sharing, and coordinated emergency response (Marincioni et al. 2014). The development of common protocols and methodologies will be critical to improving interregional and transnational cooperation, ensuring better preparedness and coordinated action to address transboundary risks. In essence, addressing climate change requires multilevel governance that operates at multiple scales and requires collaborative efforts that transcend national boundaries. It involves a coordinated interplay between local, regional, and international entities, focusing on proactive measures, information sharing, and cohesive response strategies to build resilience in the face of climate-related disasters.

Multilevel governance recognizes that issues like climate change require a synchronized effort involving multiple stakeholders. Policymakers who shape regulations and funding, scientists who provide critical insights and solutions, and citizens who are directly affected have a key role to play, especially in adaptation efforts.

First, as mentioned, governments and policymakers at all levels should be involved. Policymakers at the international and national levels are responsible for enacting laws, setting targets, and providing funding for climate initiatives (Pangalos 2023). They allocate resources, direct funding, and create policies that encourage climate-friendly practices (Rayner 2023). Regional and local governments enact laws and regulations that incentivize the implementation of adaptation strategies and emphasize engagement with diverse stakeholders (Gioia et al. 2020). Indeed, these entities often have direct knowledge of local needs and can implement tailored approaches (Bhave et al. 2013; Čehić Marić et al. 2023).

Second, scientists should be involved to provide essential data-based knowledge on the impacts of climate change and to suggest viable adaptation options (Bhave et al. 2013). Their expertise supports policymakers in making informed decisions at different scales (Bormann et al. 2012). Moreover, they contribute to both "top-down" approaches (collecting and evaluating established strategies) and "bottom-up" approaches (involving stakeholders in defining effective strategies for their areas) (Bormann et al. 2012). For this reason, social scientists and researchers are urged to engage practically in policy development to contribute directly to strategies and to analyze their impact on governance and policy (Rayner 2023).

Third, the active participation of citizens is crucial. They possess invaluable local knowledge and preferences that influence the effectiveness and acceptability of adaptation measures (Bormann et al. 2012). Promoting inclusive democracy in decision-making processes ensures that strategies are responsive to local needs and reflect their concerns and aspirations (Bhave et al. 2013). In addition, citizens ultimately implement policies and provide critical feedback on the effectiveness and acceptance of adopted strategies (Bormann et al. 2012).

This multi-stakeholder approach, integrating top-down policies with bottom-up involvement, forms the bedrock for effective climate adaptation efforts.

In Europe, the response to climate change has been multifaceted, initially focusing on reducing greenhouse gas emissions and later expanding to include adaptation measures (Biesbroek et al. 2010). The European Union has played a key role in setting targets and strategies for both mitigation and adaptation, recognizing the need for coordinated efforts across different sectors. In the European context, the development of the concept of European solidarity represents a paradigm shift in addressing climate change. The principle of subsidiarity is defined in Article 5(3) of the Treaty on European Union (1992). It aims to ensure that decisions are taken at the level closest to the citizen, by verifying that the action to be taken at the European Union level is justified in relation to the possibilities offered by action at national, regional or local level. More specifically, it is the principle that the EU should intervene only in areas where it has exclusive competence, unless its action is considered more effective than that taken at national, regional, or local level.

The recent EU Adaptation Strategy (2021) and the European Green Deal emphasize a comprehensive and transformative approach to adaptation, aiming for systemic changes in socio-ecological systems and "just" resilience to climate impacts, as well as climate neutrality (Rayner 2023). Furthermore, the 2021 European Climate Law emphasized the need for both the EU and its member states to advance adaptation efforts (European Regulation 2021). In addition, transnational networks of local authorities, such as the Covenant of Mayors for Climate and Energy, continue to advocate for action at the EU level to ensure better support from national governments for their initiatives.

However, challenges remain due to insufficient knowledge, political support, and funding at the national and subnational levels, hindering effective policy implementation (Juhola and Westerhoff 2011; Rayner 2023). Local leaders often lack the authority to take essential actions, while municipal-level policies remain constrained by sectoral boundaries and lack meaningful engagement with vulnerable stakeholders. Despite proposals for a more stringent regulatory approach, the Commission has avoided confrontation with national governments in its latest strategy, opting instead for the adoption of the EU Mission: Adapting to Climate Change (European Commission 2023). This initiative aims to support 150 European regions, cities, and local authorities in their efforts to build resilience to the impacts of climate change by 2030. Nevertheless, some critical issues emerge, such as the understanding of the interpretation of "just" resilience and its implications (Lager et al. 2023), the ability to promote transformative concepts in the midst of resistance (De Rosa et al. 2022), the effectiveness of coordinating regional initiatives across borders (Rayner 2023), and the coherence between economic, social, and environmental policies in the same localities (Göpfert et al. 2019). To overcome this, Lenaerts et al. (2022) propose a three-level governance for adaptation measures (European, national, and subnational), emphasizing intensive cooperation, binding adaptation plans and an EU-level insurance scheme linked to national targets. In addition, they propose increased funding for vulnerable regions through EU policies and financial instruments to protect critical infrastructure. In fact, the current EU strategy lacks binding and measurable targets, leading to weak policy implementation due to lack of awareness, political prioritization, and funding (Lenaerts et al. 2022).

This book is a comprehensive effort that exceeds the mere compilation of adaptation strategies for the Adriatic region. It seeks to provide a methodological framework that effectively addresses the implementation challenges inherent in existing adaptation policies. The scope encompasses the complex interrelationships between European, national, and regional levels of governance in the field of adaptation, as well as the promotion of cooperation between key stakeholders, namely policymakers, scientists, and citizens.

Essentially, the work examines the existing and prevailing strategies at multiple levels of governance and analyzes their nuances, commonalities, and divergences. It builds upon the urgency imposed by climate change, which has significantly increased natural hazards and consequent risks, with a particular focus on marine and coastal areas along the Adriatic Sea. This urgency is forcing local governments and authorities to face the imperative task of identifying, preparing for and reducing the impacts of climate change. The RESPONSe project, developed in the Adriatic Basin under the EU Interreg Italy-Croatia initiative, embodies this urgency. Its main objectives include improving local climate knowledge, involving local authorities and stakeholders in planning adaptation and mitigation strategies, and providing technical support for local strategy development.

After exploring the observed and projected climate risks affecting the Adriatic Basin, the reader is immersed in the landscape of current European, national, and regional policies. In this phase, the analysis will focus on both adaptation and mitigation efforts, in order to best establish the overall context and potential linkages. From this perspective, it will be possible to understand, from a scientific point of view, whether the current strategies really address the most urgent climate-related challenges or only those imposed by the higher level of governance. The final analysis is dedicated to the creation of an open database of adaptation measures already successfully implemented in the Adriatic region and beyond, in order to pave the way for the implementation of adaptation strategies with actions tailored to the specific local contexts. This analysis will capitalize on the outcomes of several projects and initiatives aimed at focusing on climate change adaptation and strengthened resilience of Adriatic local and regional communities, such as the Interreg Adri-Adapt project (Magni et al. 2021), and extending up to national, European, and international databases (Aguiar et al. 2018; Reckien et al. 2018; Pietrapertosa et al. 2019).

In the end, such a methodological framework will help to bridge the gaps in policy implementation and to promote robust, effective, and collaborative climate adaptation strategies for the Adriatic coastal zone.

References

Adger WN, Arnell NW, Tompkins EL (2005) Successful adaptation to climate change across scales. Glob Environ Change 15(2):77–86

Aguiar FC, Bentz J, Silva JMN, Fonseca AL, Swart R, Duarte Santos F, Penha-Lopes G (2018) Adaptation to climate change at local level in Europe: an overview. Environ Sci Policy 86:38–63. https://doi.org/10.1016/j.envsci.2018.04.010

Arnell NW (2010) Adapting to climate change: an evolving research programme. Clim Change 100(1):107–111

Bhave AG, Mishra A, Raghuwanshi NS (2013) A combined bottom-up and top-down approach for assessment of climate change adaptation options. J Hydrol 518:150–161. https://doi.org/10.1016/j.jhydrol.2013.08.039

Biesbroek GR, Swart RJ, van der Knaap WGM (2009) The mitigation-adaptation dichotomy and the role of spatial planning. Habitat Int 33(3):230–237

Biesbroek GR, Swart RJ, Carter TR, Cowan C, Henrichs T, Mela H, Morecroft MD, Rey D (2010) Europe adapts to climate change: comparing national adaptation strategies. Glob Environ Change 20(3):440–450. https://doi.org/10.1016/j.gloenvcha.2010.03.005

Bormann H, Ahlhorn F, Klenke T (2012) Adaptation of water management to regional climate change in a coastal region—hydrological change vs. community perception and strategies. J Hydrol 454–455:64–75. https://doi.org/10.1016/j.jhydrol.2012.05.063

Čehić Marić A, Čop T, Oplanić M, Goreta Ban S, Njavro M (2023) Adaptation to climate change in Adriatic Croatia—the view of policymakers. Sustainability 15:7085. https://doi.org/10.3390/su15097085

De Rosa SP, de Moor J, Dabaieh M (2022) Vulnerability and activism in urban climate politics: an actor-centered approach to transformational adaptation in Malmö (Sweden). Cities 130:103848

European Commission (2000) EU policies and measures to reduce greenhouse gas emissions: towards a European climate change programme (ECCP). COM(2000) 88 final

European Commission (2003) Establishing a scheme for greenhouse gas emission allowance trading within the community and amending Council Directive 96/61/EC, Directive 2003/87/EC of the European Parliament and of the Council, dated 13 October 2003

European Commission (2009) White paper—adapting to climate change: towards a European framework for action. COM/2009/0147 final

European Commission (2013a) An EU strategy on adaptation to climate change. COM(2013) 216 final

European Commission (2013b) Green paper on the insurance of natural and man-made disasters. COM(2013) 213 final

European Commission (2014) A policy framework for climate and energy from 2020 to 2030. COM(2014) 0015

European Commission (2023) EU mission: adaptation to climate change. https://research-and-innovation.ec.europa.eu/funding/funding-opportunities/funding-programmes-and-open-calls/horizon-europe/eu-missions-horizon-europe/adaptation-climate-change_en

European Environment Agency—EEA (2023) FAQ: what is the difference between adaptation and mitigation? https://www.eea.europa.eu/help/faq/what-is-the-difference-between

European Regulation (2021) 2021/1119 of the European Parliament and of the Council of 30 June 2021 establishing the framework for achieving climate neutrality and amending Regulations (EC) No 401/2009 and (EU) 2018/1999 ('European Climate Law')

Füssel HM (2007) Adaptation planning for climate change: concepts, assessment approaches, and key lessons. Sustain Sci 2:265–275. https://doi.org/10.1007/s11625-007-0032-y

Gioia E, Colocci A, Marchetti N (2020) Strategie di adattamento e mitigazione ai cambiamenti climatici: analisi delle frontiere in Adriatico. In: Zilli S, Modaffari G (eds) Confin(at)i/Bound(aries), Società di Studi Geografici. Memorie geografiche, NS 18, pp 831–843

Göpfert C, Wamsler C, Lang W (2019) A framework for the joint institutionalization of climate change mitigation and adaptation in city administrations. Mitig Adapt Strateg Glob Change 24:1–21

Intergovernmental Panel on Climate Change—IPCC (2001) Climate change 2001: impacts, adaptation and vulnerability. Cambridge University Press

Intergovernmental Panel on Climate Change—IPCC (2018) Global warming of 1.5°C. Cambridge University Press. https://doi.org/10.1017/9781009157940

Juhola S, Westerhoff L (2011) Challenges of adaptation to climate change across multiple scales: a case study of network governance in two European countries. Environ Sci Policy. https://doi.org/10.1016/j.envsci.2010.12.006

Lager F, Coninx I, Breil M, Bakhtaoui I, Pedersen AB, Mattern K, van den Berg H, Sini E, Galluccio G, Klein R, Vierikko K (2023) Just resilience for Europe: towards measuring justice in climate change adaptation. ETC CA. https://doi.org/10.25424/CMCC-BATP-3M95

Lenaerts K, Tagliapietra S, Wolff GB (2022) How can the European Union adapt to climate change? Intereconomics 57:314–321. https://doi.org/10.1007/s10272-022-1071-4

Magni F, Carraretto G, Musco F (2021) Towards the Adriatic climate-transition: a methodological framework for adaptation spatial planning processes. Anteferma Edizioni

Marincioni F, Toseroni F, Appignanesi L (2014) Macroregione Adriatico-Ionica: Aumentare la connessione per diminuire i rischi. In: Pugnaloni F (ed) Architecture as heritage: Percorsi adriatici di progetto nel paesaggio degli insediamenti minori. Il lavoro editoriale

Mathison C, Wiltshire A, Dimri AP, Falloon P, Jacob D, Kumar P, Moors E, Ridley J, Siderius C, Stoffel M, Yasunari T (2013) Regional projections of North Indian climate for adaptation studies. Sci Total Environ 468–469:S4–S17. https://doi.org/10.1016/j.scitotenv.2012.04.066

Nalau J, Preston BL, Maloney MC (2015) Is adaptation a local responsibility? Environ Sci Policy 48:89–98. https://doi.org/10.1016/j.envsci.2014.12.011

Pangalos C (2023) Tackling climate change through multi-level governance, systems thinking and policy coherence. https://www.unssc.org/news-and-insights/blog/tackling-climate-change-through-multi-level-governance-systems-thinking-and

Pietrapertosa F, Salvia M, De Gregorio Hurtado S, D'Alonzo V, Church JM, Geneletti D, Musco F, Reckien D (2019) Urban climate change mitigation and adaptation planning: are Italian cities ready? Cities 91:93–105. https://doi.org/10.1016/j.cities.2018.11.009

Rayner T (2023) Adaptation to climate change: EU policy on a mission towards transformation? npj Clim Action 2:36. https://doi.org/10.1038/s44168-023-00068-z

Reckien D, Salvia M, Heidrich O, Church JM, Pietrapertosa F, De Gregorio-Hurtado S, D'Alonzo V, Foley A, Simoes SGS, Lorencová EK, Orru H, Orru K, Wejs A, Flacke J, Olazabal M, Geneletti D, Feliu E, Vasilie S, Nador C, Dawson RJ (2018) How are European cities planning to respond to climate change? Analysis of local climate plans from 885 urban areas in the EU-28. J Clean Prod 191:207–219. https://doi.org/10.1016/j.jclepro.2018.03.220

Rogelj J, Popp A, Calvin K, Luderer G, Emmerling J, Gernaat D, Fujimori S, Strefler J, Hasegawa T, Marangoni G, Krey V, Kriegler E, Riahi K, Van Vuuren D, Doelman J, Drouet L, Edmonds J, Fricko O, Harmsen M, Havlík P, Humpenöder F, Stehfest E, Tavoni M (2018) Scenarios towards limiting global-mean temperature increase below 1.5 °C. Nat Clim Change 8(3):325–332. https://doi.org/10.1038/s41558-018-0091-3

Smith B, Pilifosova O (2001) Adaptation to climate change in the context of sustainable development and equity. In: Third assessment report

United Nations General Assembly—UNGA (1994) United Nations framework convention on climate change: resolution/adopted by the General Assembly. A/RES/48/189. https://www.refworld.org/docid/3b00f2770.html

Chapter 2
Understanding Climate Risks in the Adriatic Basin

Since recent decades, a disaster in one EU Member State is no longer an isolated event. It is now recognized as a challenge that affects all member states and requires the mobilization of the entire European apparatus. A shift from the disjointed dispatch of national resources to the affected country to a more coordinated and structured approach led directly by the European Union, has been acknowledged. This shift is based on the principle of subsidiarity among European nations, a principle that decisions are best made at the most local level possible. This development represents a shift from mere acts of solidarity to the recognition of shared responsibility and the activation of a collective response mechanism across the European landscape (Marincioni et al. 2014).

Indeed, at the international level, the United Nations Disaster Risk Reduction (DRR) policy has notably shifted its focus from post-event relief to a greater emphasis on proactive measures taken before such events occur (UNISDR 2015). The goal is to reduce risk, defined as the potential economic loss resulting from the exposure of people and their infrastructure to extreme natural events. This shift takes place within a broader framework that emphasizes the territorial resilience of communities and systems.

According to the European Environment Agency's CATDAT disaster database (EEA 2023), between 1980 and 2022, meteorological (e.g., storms, including lightning and hail, together with mass movements), hydrological (e.g., floods) and climatological (e.g., heat waves, cold waves, droughts, forest fires) events have caused an estimated €650 billion in economic losses in the EU Member States, of which €59.4 billion in 2021 and €52.3 billion in 2022 (Fig. 2.1). Hydrological hazards account for nearly 43% and meteorological hazards for about 29% of the total. For climatological hazards, heat waves cause about 20% of the total losses, while the remaining 8% are caused by droughts, forest fires, and cold waves together. The EU countries with the highest economic losses in the period 1980–2022 were Germany, France, and Italy. The most expensive events include the 2021 flood in Germany and

E. Gioia, *Climate Change Adaptation in the Adriatic Coastal Areas*, Disaster Risk Reduction, https://doi.org/10.1007/978-981-97-6655-0_2

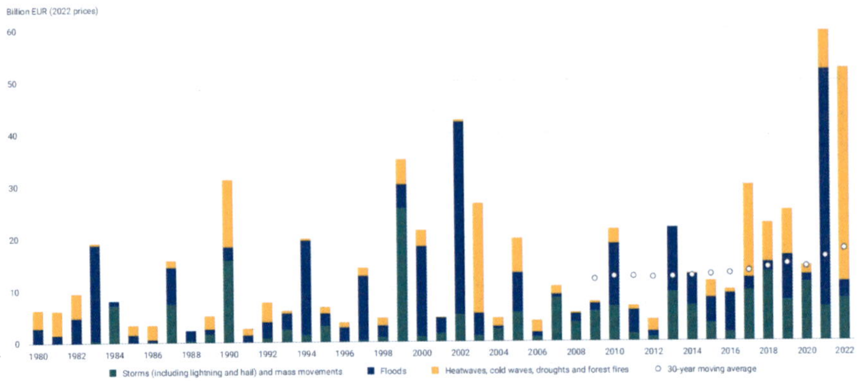

Fig. 2.1 Annual economic losses caused by weather—and climate—related extreme events in the EU Member States. *Source* European Environmental Agency (2023)

Belgium (€44 billion), the 2022 combined drought and heat events across the continent (€40 billion), the 2002 flood in central Europe (€34 billion), the 1999 Lothar storm in western Europe (€17 billion), the 2003 drought and heatwave across the EU (€17 billion), and the 2000 flood in France and Italy (€14 billion).

These events have far exceeded the administrative boundaries of individual countries and have had a significant impact on the economies, transport networks, and environment of the entire EU area. They are a stark reminder that disasters and environmental crises know no borders. They demonstrate the interconnectedness of nations within the EU, where an incident in one region can reverberate across the European landscape, affecting several countries and their citizens. These incidents highlight the urgent need for collaborative, cross-border strategies, and international cooperation to effectively respond to and mitigate the consequences of such crises (Marincioni et al. 2014).

Over the years, there has been an increase in average annual economic losses attributed to climate-related events, with the highest recorded losses occurring in recent years (Seneviratne et al. 2021). Projections from the Intergovernmental Panel on Climate Change (IPCC) suggest that these extreme events will intensify, potentially leading to systemic failures across Europe and resulting in greater economic losses (Bednar-Friedl et al. 2022; CMCC 2021). Consequently, it seems improbable that economic losses will decrease by 2030.

Given the recognition that many major disasters are linked to global climate change and the often vast scale of these phenomena, policies that are limited to purely national or territorial dimensions are no longer effective. It is now clear that addressing these issues requires action at multiple scales, from the local to the global. Indeed, cooperation at the local level and a broader international perspective are both essential to effectively address these challenges.

In this sense, the Adriatic Basin, which includes the Adriatic Sea and 6 adjacent countries, that is Italy, Slovenia, Croatia, Bosnia and Herzegovina, Montenegro, and Albania, stands out as a significant geographical area (Fig. 2.2). It is characterized

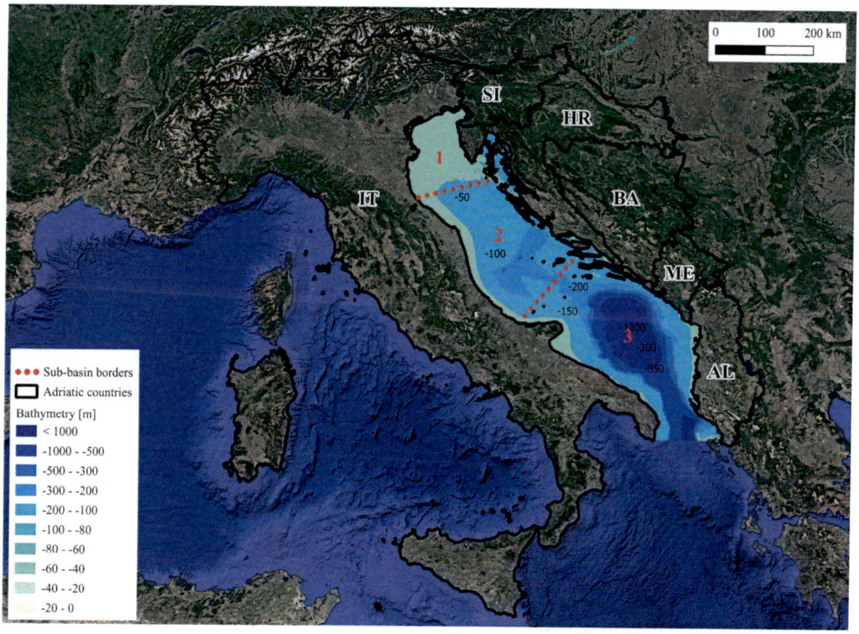

Fig. 2.2 Map of the Adriatic Basin, showing the extent and bathymetry of the Adriatic Sea, and the national borders of Italy (IT), Slovenia (SI), Croatia (HR), Bosnia and Herzegovina (BA), Montenegro (ME) and Albania (AL). Boundaries of the sea are approximated and divided in three sub-basins, according to the bathymetry: 1. Northern Adriatic; 2. Central Adriatic; 3. Southern Adriatic. *Source* Google Earth basemap; bathymetric details from the GEBCO grid (2023)

by peculiar geomorphological aspects. The mean depth is − 253 m, with a median of − 100 m and a maximum of − 1244 m (Vrdoljak et al. 2021). The Adriatic Sea can be divided into northern, central and southern sub-basins, which have different depth ranges and seafloor topography (Fig. 2.2). The Northern Adriatic sub-basin is the shallowest part of the Adriatic Sea, accounting for only 4% of the total basin volume, with an average depth of − 43 m. The central Adriatic sub-basin is a transitional zone between the shallow northern part and the deepest southern part, which contains 14% of the total volume of the Adriatic Sea with an average depth of − 110 m. The southern Adriatic sub-basin contains more than 80% of the volume of the Adriatic with an average depth of − 498 m, but the mean depth is only − 295 m due to the terrain configuration. Therefore, the Adriatic is a shallow sea that tends to be quickly filled by sediments resulting from erosion. In fact, it is destined to disappear in the near geological future.

Throughout history, this sea has served not as a barrier but as a cohesive force among the populations living along its shores (Anselmi 2004). Various metaphors have been used to describe the Adriatic: a "liquid plain" surrounded by inhabited mountains and a north–south road central to the position of Venice (Braudel 1976), and even a "highway", emphasizing its importance compared to land routes (Di

Venosa 2001). In fact, throughout history, sea routes were faster and cheaper than land routes until the advent of railroads (Mollat du Jourdin 1993). From the Middle Ages to the nineteenth century, the Adriatic Sea played a bridging and intermediary role not only economically, but also politically and culturally. Geographically, its position facilitated this role, especially in the medieval and modern periods, acting as a link between the east of the Roman Empire and Carolingian Europe (Anselmi 1991). The Adriatic became a meeting point for the Byzantine, Frankish, and Arab worlds, especially under the rule of the Republic of Venice (Moroni 2014). Commercial, cultural, and artistic exchanges flourished, creating a complex web of relationships and trade treaties among Adriatic cities. Scholars such as Fernand Braudel (1976) and Jean-Claude Hocquet (1979) emphasized the economic and cultural unity of the Adriatic, describing it as an "Adriatic state" before the emergence of modern nations. In the nineteenth and twentieth centuries, Italy's national unification triggered structural changes that allowed it to catch up with the industrialization of western Europe. Industrial developments along the Adriatic coast transformed the region, leading to economic, social, and even ecological changes (Battilani 2001; Moroni 2014). Later, conflicts in the Balkans, fueled by the economic interests of Western powers, necessitated renewed cooperation. The eastward expansion of the European Union and the formation of a Euro-Adriatic region aimed to promote transportation infrastructure and regional development (Rosini 2001).

Today, amidst the complexities of the global market, the Adriatic Sea can remain a maritime route or evolve into the core of a macro-region that promotes cross-border cooperation and reclaims its traditional role as a hub of cultural and economic exchange (Moroni 2014). This vision requires both a long-term perspective and short- to medium-term strategies to rebuild economic and cultural relations within Europe, potentially serving as a new frontier between wealthy and less wealthy countries. Ultimately, the Adriatic, with its historical centrality between east and west, could regain its pivotal role by becoming the fulcrum of a new Euro-region, restoring its ancient function as a gateway between Europe and the east, as well as North African countries (Moroni 2014).

The destiny for the Adriatic populations is also to be united by the same hazards and risks associated with the physical characteristics of the area. Geological, geomorphological and climatic features form a common denominator that renders the Adriatic territories potentially susceptible to the same hazards.

From a geological perspective, the area, which includes Italy, Croatia, Albania, and northern Greece, contains six fault lines, with the Aegean Sea posing significant geological hazards. One prominent fault, the Gloria, extends from Gibraltar and transforms into a deformation zone that traverses the Mediterranean Sea, the Apennines, and the Croatian, Albanian, and Greek coastlines (Baptista et al. 2016). The interaction of the African and European plates, together with the subduction of the Adriatic microplate beneath the Italian peninsula, highlights the prevailing seismic hazards in the Adriatic basin. Over the past 2500 years, Italy has experienced approximately 560 earthquakes, ranging from strong to catastrophic, with magnitudes between 8 and 10 on the Mercalli scale, occurring approximately every four and a half years (Marincioni et al. 2014). In addition, this geological setting poses a risk of tsunamis.

In the last millennium, several tsunamis have struck the Adriatic, causing significant events, such as sea level rise and flooding, in the northern Adriatic in 1511 and 1830, central Adriatic in 1672 and 1930, southern Adriatic in 1627 and 1920, and most recently an event in 1979 that affected Montenegro (Marincioni et al. 2014; Pranavam Ayyappan Pillai et al. 2022).

In terms of climatic characteristics, the Adriatic coast is a representative example of a coastal zone subject to a variety of significant and rapidly evolving meteorological, hydrogeological, and hydraulic hazards that recur along the Mediterranean coastline (Simeoni and Corbau 2009). The most frequent manifestations of these phenomena for the population living in coastal areas are sea level inundation, aggravated by natural or anthropogenic subsidence in the northern Adriatic, but also storm surge flooding, coastal erosion, wind waves, and meteotsunamis (Torresan et al. 2012; Gallina et al. 2020; Maramai et al. 2022; Pranavam Ayyappan Pillai et al. 2022). Other frequent phenomena, even in inland areas, are landslides and avalanches, storms, cyclones, heat waves, and droughts (Gioia et al. 2021; Cavaleri et al. 2019; Bonaldo et al. 2023; Toomey et al. 2022; Ricchi et al. 2023).

These considerations are compounded by climate change, which is impacting the Adriatic Basin with phenomena such as extreme weather events, sea level rise that influences freshwater availability through salinization, and increased fire risks associated with droughts and heat waves, affecting various aspects of human life and local economies. In particular, the Interreg Italy-Croatia AdriAdapt project (https://adriadapt.eu/) has produced extensive climate data and projections from 1986 to 2100 focused on the Adriatic coastal regions of Italy and Croatia. These projections are based on two emission scenarios: a moderate one (RCP4.5), reflecting significant mitigation efforts by society, and a high emission scenario (RCP8.5), representing a world without mitigation policies. The projections highlight shifts in several climate indices, such as extreme temperature and precipitation, as well as combined indices that affect both population and land. Some of the key findings of the analysis include: (i) an increase in seasonal and annual mean air temperature by 2100, especially in summer; (ii) a shift toward a warmer temperature distribution across all seasons, including maximum and minimum values; (iii) a decrease in the number of frost days in winter, spring, and fall, with some areas experiencing the disappearance of frost days, especially under RCP8.5 toward the end of the century; (iv) a significant increase in the duration of heat waves and tropical nights, particularly affecting urban areas in summer; (v) variability in precipitation projections across seasons and scenarios, with evidence of reduced summer precipitation and increased intense winter precipitation. Such results reflect the climate risks that both Italy and Croatia face. According to the World Bank (2021), Croatia is expected to become hotter and drier, especially in summer; future precipitation trends are projected to decrease steadily over the century, especially in summer; and highly damaging floods, including coastal and riverine floods, are expected to occur at least once every 10 years. According to Euro-Mediterranean Center on Climate Change, climate change in Italy is projected to bring about a + 5 °C rise in average temperature by 2100, altering precipitation patterns with decreased annual precipitation, increased intensity on rainy days, more hot and dry days, and adverse effects on the marine environment, impacting coastal ecosystem services and the socio-economic system (Spano et al. 2020).

In addition, Tojčić et al. (2023) found several ongoing processes in the Adriatic Sea: (a) extensive warming of the surface atmosphere during summer, which increases both temperature and salinity, (b) shallowing of the advection of the saline and warmer Levantine intermediate water inflow into the Adriatic, (c) decrease of the Adriatic deep water outflow, which is now more restricted to the bottom, Adriatic-Ionian thermohaline circulation, (d) warming of the near-bottom waters, especially in the central and northern Adriatic, and (e) shrinking and weakening of the southern Adriatic cyclonic gyre, which is a large system of counterclockwise circulating surface ocean currents. Such processes affect the exchange of water, its oxygenation, and thus represent a factor in the multiplication of pollutants. The risk in a future scenario is the transformation of the Adriatic Sea into a kind of salt marsh (Marincioni et al. 2014). This scenario is even more serious when considering the enormous amount of waste that lies at the bottom of the Adriatic Sea. According to experts from the Polytechnic University of Marche, Department of Life and Environmental Sciences, the Adriatic Sea is predicted to be a preferential area of plastic accumulation within the Mediterranean Sea (Avio et al. 2020). The area is under strong anthropic pressure due to intense coastal urbanization. In fact, several anthropogenic activities coexist, such as tourism, heavy maritime traffic from commercial and tourist vessels, offshore activities, ferry boats, and trawling (Mazzoli and Gorbi 2021). In addition, a large number of rivers discharge into the basin with a significant impact on the circulation. In particular, the Adriatic Sea collects one third of the freshwater flowing into the Mediterranean Sea, mainly via the Po River (Gajšt et al. 2016).

As a result, these hazards and the potential economic losses resulting from societal exposure to these hazards are also strongly influenced by human activities. Population density, increasing urbanization, abandonment of mountainous areas, illegal construction, continued deforestation, use of unsustainable agricultural practices, and lack of maintenance of land and waterways have exacerbated disturbances and highlighted the fragility of the land. These human activities increase exposure to natural phenomena and, therefore, risk.

Given the vulnerability of the Adriatic system, addressing climate change issues in a cross-border perspective becomes crucial for this region of Europe. Working on Disaster Risk Reduction at the Adriatic Basin level presents both opportunities and challenges (Marincioni et al. 2014). Environmental disasters, like those deriving from climate change, transcend national borders, necessitating an international scale for better defining hazards and risk scenarios. However, linguistic diversity, varied procedures, organizations, and authorities within the Basin pose obstacles to coordinated risk prevention and mitigation efforts. Overcoming national boundaries requires enhanced communication and collaboration among agencies. Developing shared communication protocols and data exchange is critical for improved interregional and transnational cooperation in Disaster Risk Reduction, ultimately leading to better harmonization of adaptation strategies to ongoing climate changes. A macro-regional approach aims to ensure greater comparability among existing national and regional methods, utilizing a wealth of local knowledge in a systemic manner.

References

Anselmi S (1991) Adriatico. Studi di storia Secoli XIV–XIX. Ancona

Anselmi S (2004) I mediterranei e il Mediterraneo. Prop Ricerche 52:13–27

Avio CG, Pittura L, d'Errico G, Abel S, Amorello S, Marino G, Gorbi S, Regoli F (2020) Distribution and characterization of microplastic particles and textile microfibers in Adriatic food webs: general insights for biomonitoring strategies. Environ Pollut 258:113766

Baptista MA, Miranda JM, Batlló J, Lisboa F, Luis J, Macià R (2016) New study on the 1941 Gloria Fault earthquake and tsunami. Nat Hazards Earth Syst Sci 16(8):1967–1977

Battilani P (2001) Vacanze di pochi, vacanze di tutti. L'evoluzione del turismo europeo. Bologna

Bednar-Friedl B, Biesbroek R, Schmidt DN, Alexander P, Børsheim KY, Camicer J, Georgopoulou E, Haasnoot M, Le Cozannet G, Lionello P, Lipka O, Möllmann C, Muccione V, Mustonen T, Piepenburg D, Whitmarsh L (2022) Europe. In: Pörtner H-O, Roberts DC, Tignor MMB et al (eds) Climate change 2022: impacts, adaptation and vulnerability. Contribution of working group II to the sixth assessment report of the intergovernmental panel on climate change. Cambridge University Press

Bonaldo D, Bellafiore D, Ferrarin C et al (2023) The summer 2022 drought: a taste of future climate for the Po valley (Italy)? Reg Environ Change 23(1). https://doi.org/10.1007/s10113-022-02004-z

Braudel F (1976) Civiltà e imperi del Mediterraneo nell'età di Filippo II, 2 vol. Turin

Cavaleri L, Bajo M, Barbariol F, Bastianini M, Benetazzo A, Bertotti L, Chiggiato J, Davolio S, Ferrarin C, Magnusson L (2019) The October 29, 2018 storm in Northern Italy—an exceptional event and its modeling. Prog Oceanogr 178:102178

Di Venosa M (2001) Le autostrade in mare. Adriatico 2:23–26

Euro-Mediterranean Center on Climate Change—CMCC (2021) G20 climate risk atlas: impacts, policy, economics. European Union. https://files.cmcc.it/g20climaterisks/Eu27.pdf

European Environment Agency—EEA (2023) Economic losses from weather- and climate-related extremes in Europe. https://www.eea.europa.eu/en/analysis/indicators/economic-losses-from-climate-related

Gajšt T, Bizjak T, Palatinus A, Liubartseva S, Kržan A (2016) Sea surface microplastics in the Slovenian part of the Northern Adriatic. Mar Pollut Bull 113(1–2):392–399

Gallina V, Torresan S, Zabeo A, Critto A, Glade T, Marcomini A (2020) A multi-risk methodology for the assessment of climate change impacts in coastal zones. Sustainability 12(9):3697. https://doi.org/10.3390/su12093697

General Bathymetric Chart of the Ocean—GEBCO (2023) GEBCO 2023 digital bathymetry model. https://download.gebco.net/

Gioia E, Marincioni F, Ferretti M (2021) Modeling landslide hazard in the Esino River Valley (Central Italy). In: Geographies of the anthropocenes. Il Sileno Edizioni, Lago (CS)

Hocquet JC (1979) Il sale e la fortuna di Venezia. Rome

Maramai A, Brizuela B, Graziani L (2022) A database for tsunamis and meteotsunamis in the Adriatic Sea. Appl Sci 12:5577. https://doi.org/10.3390/app12115577

Marincioni F, Toseroni F, Appignanesi L (2014) Macroregione Adriatico-Ionica: Aumentare la connessione per diminuire i rischi. In: Pugnaloni F (ed) Architecture as heritage: Percorsi adriatici di progetto nel paesaggio degli insediamenti minori. Il lavoro editoriale

Mazzoli C, Gorbi S (2021) Plastic litter in the Adriatic Basin. Greenpeace report. https://www.greenpeace.org/static/planet4-italy-stateless/2021/06/98c3eab4-report-plastic-litter-in-the-adriatic-basin-2021.pdf

Mollat du Jourdin M (1993) L'Europa e il mare. Rome-Bari
Moroni M (2014) Evoluzione storica dei flussi della koinè Adriatica. In: Pugnaloni F (ed) Architecture as heritage: Percorsi adriatici di progetto nel paesaggio degli insediamenti minori. Il lavoro editoriale
Pranavam Ayyappan Pillai U, Pinardi N, Federico I, Causio S, Trotta F, Unguendoli S, Valentini A (2022) Wind-wave characteristics and extremes along the Emilia-Romagna coast. Nat Hazards Earth Syst Sci 22:3413–3433. https://doi.org/10.5194/nhess-22-3413-2022
Ricchi A, Sangelantoni L, Redaelli G, Mazzarella V, Montopoli M, Miglietta MM, Tiesi A, Mazzà S, Rotunno R, Ferretti R (2023) Impact of the SST and topography on the development of a large-hail storm event, on the Adriatic Sea. Atmos Res 296:107078. https://doi.org/10.1016/j.atmosres.2023.107078
Rosini R (2001) L'Adriatico come rete di infrastrutture e come area-mercato dei trasporti internazionali. Adriatico 1:31–41
Seneviratne SI, Zhang X, Adnan M, Badi W, Dereczynski C, Di Luca A, Ghosh S, Iskandar I, Kossin J, Lewis S, Otto F, Pinto I, Satoh M, Vicente-Serrano SM, Wehner M, Zhou B (2021) Weather and climate extreme events in a changing climate. In: Climate change 2021: the physical science basis. Contribution of working group I to the sixth assessment report of the intergovernmental panel on climate change. Cambridge University Press. https://doi.org/10.1017/9781009157896.013
Simeoni U, Corbau C (2009) A review of the Delta Po evolution (Italy) related to climatic changes and human impacts. Geomorphology 107:64–71
Spano D, Mereu V, Bacciu V, Marras S, Trabucco A, Adinolfi M, Barbato G, Bosello F, Breil M, Chiriacò MV, Coppini G, Essenfelder A, Galluccio G, Lovato T, Marzi S, Masina S, Mercogliano P, Mysiak J, Noce S, Pal J, Reder A, Rianna G, Rizzo A, Santini M, Sini E, Staccione A, Villani V, Zavatarelli M (2020) Analisi del rischio: I cambiamenti climatici in Italia. https://doi.org/10.25424/CMCC/ANALISI_DEL_RISCHIO
Tojčić I, Denamiel C, Vilibić I (2023) Kilometer-scale trends and variability of the Adriatic present climate (1987–2017). Clim Dyn 61:2521–2545. https://doi.org/10.1007/s00382-023-06700-2
Toomey T, Amores A, Marcos M, Orfila A, Romero R (2022) Coastal hazards of tropical-like cyclones over the Mediterranean Sea. J Geophys Res Oceans 127:e2021JC017964. https://doi.org/10.1029/2021JC017964
Torresan S, Critto A, Rizzi J, Marcomini A (2012) Assessment of coastal vulnerability to climate change hazards at the regional scale: the case study of the North Adriatic Sea. Nat Hazards Earth Syst Sci 12:2347–2368. https://doi.org/10.5194/nhess-12-2347-2012
United Nations International Strategy for Disaster Reduction—UNISDR (2015) Sendai framework for disaster risk reduction 2015–2030. Geneva
Vrdoljak L, Režić M, Petričević I (2021) Bathymetric and geological properties of the Adriatic Sea. Rud-geol-naft zb 36(2):93–107. https://doi.org/10.17794/rgn.2021.2.9
World Bank (2021) Climate risk country profile—Croatia. Accessed 8 Dec 2023. https://reliefweb.int/report/croatia/climate-risk-country-profile-croatia

Chapter 3
The Adriatic Case Studies Within the Project RESPONSe

The RESPONSe project (*Strategies to adapt to climate change in Adriatic regions*) was developed through the Interreg Italy-Croatia funding system (https://www.italy-croatia.eu/web/response), one of the European Commission's tools to support a fruitful exchange between the communities on both sides of the Adriatic basin in order to optimize resources in the search for common solutions. Specifically, the RESPONSe activities have taken place between January 2019 and April 2022, with a budget of €2.14 million.

The partnership was led by Informest, a cooperation agency based in the Friuli-Venezia Giulia Region with the aim of promoting economic development and internationalization processes. Informest, together with the Polytechnic University of Marche, the Agency for Energy Management of Friuli-Venezia Giulia (APE-FVG), the Regional Agency for Environmental Protection and Prevention of Veneto (ARPAV), the Region of Apulia formed the Italian partners. The Energy Institute Hrvoje Požar (EIHP) for Renewable Energy Sources, Energy Efficiency and Environmental Protection, the Croatian Meteorological and Hydrological Service (DHMZ) and the Institute of Oceanography and Fisheries (IOF) formed the Croatian partners.

The eight partners involved worked in close collaboration with several municipalities within six study areas, corresponding to three Italian coastal Regions (Friuli-Venezia Giulia, Marche, Apulia) and three Croatian coastal Counties (Primorje-Gorski Kotar, Šibenik-Knin, Dubrovnik-Neretva), equally distributed between the northern, central and southern Adriatic (Fig. 3.1). In Italy, the Regions correspond to the second level of Nomenclature of Territorial Units for Statistics (NUTS-2), whereas in Croatia, the Counties correspond to the third level (NUTS-3). However, this choice makes the areas more comparable.

The northernmost Italian study area is Friuli-Venezia Giulia, an autonomous region with special statute, bordering Austria to the north, Slovenia to the east, the Veneto Region to the west, and the Adriatic Sea to the south. Encompassing about 7924 km^2, is a relatively small region but rich in diversity and cultural influences due to its border position. It comprises the provinces of Udine, Pordenone,

E. Gioia, *Climate Change Adaptation in the Adriatic Coastal Areas*, Disaster Risk Reduction, https://doi.org/10.1007/978-981-97-6655-0_3

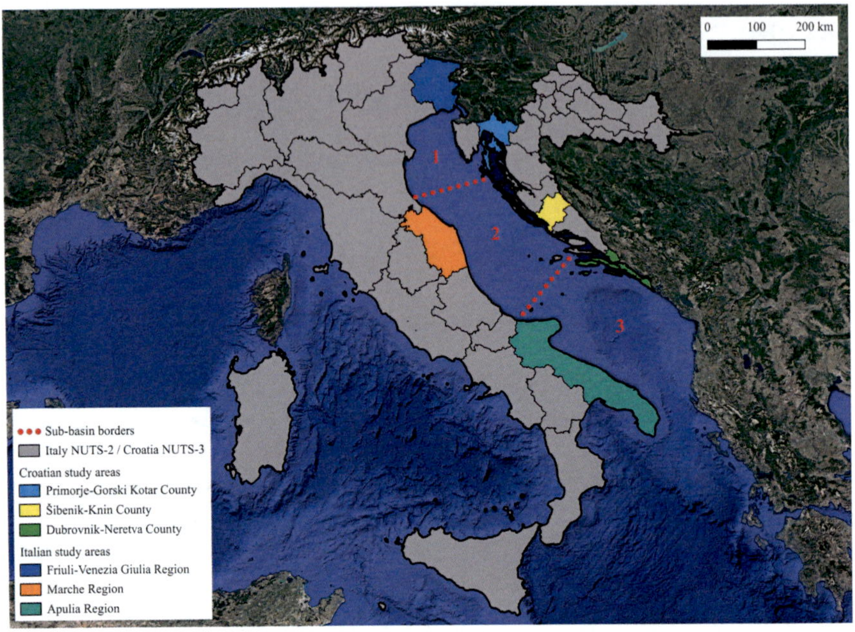

Fig. 3.1 Italian regions (NUTS-2) and Croatian Counties (NUTS-3) selected as study areas within the Adriatic Basin, each pertaining to one of the three sub-basins obtained according to the bathymetry: 1. Northern Adriatic; 2. Central Adriatic; 3. Southern Adriatic. *Source* Google Earth basemap

Gorizia, and Trieste. As of 2020, the population was recorded at 1.2 million inhabitants with a density of 159.6 persons per km^2 (Eurostat 2020). Geomorphologically, Friuli-Venezia Giulia encompasses a variety of landscapes. The Alps characterize the northern part, with rugged mountains and picturesque valleys. Toward the south, the terrain gradually descends into hills and plains, eventually meeting the Adriatic coast. The region is divided by the Tagliamento River valley and is characterized by a southeastern extension of the territory of Trieste. This diversity contributes to the region's economic activities, including agriculture, tourism, and industry. The climate in Friuli-Venezia Giulia is diverse as well. In the coastal areas, the climate tends to be more Mediterranean, characterized by mild winters and warm summers. Moving toward the inland areas and higher altitudes, the climate becomes more continental, experiencing colder winters with significant snowfall in the mountainous regions. Seismic activity is notable due to its location near the convergence of the Eurasian and African tectonic plates. Indeed, part of the territory is classified in "Zone 1" the most dangerous area, where major earthquakes may occur (DPC 2023). This is underscored by a devastating earthquake in May 1976 (Mw 6.45) that claimed over 1000 lives (Slejko et al. 2019; Rovida et al. 2022). Friuli-Venezia Giulia is also exposed to extreme climatic events (ARPA FVG 2023). In particular, a rise in sea levels is expected in the coming years, which will affect the lagoons of Grado and

Marano, changing their morphology and increasing the risk of flooding in coastal areas. Furthermore, an increase in forest fires is expected due to the drought that favors the accumulation of dry plant material. The energy consumption of buildings, especially for air conditioning in summer, is also expected to increase due to climate change.

The central Italian region of Marche lies between the Apennine Mountains and the Adriatic Sea. Covering approximately 9344 km^2, it is bordered by Emilia-Romagna and the Republic of San Marino to the north, Tuscany and Umbria to the west, and Lazio and Abruzzo to the south. Comprising five provinces, Ancona, Ascoli Piceno, Fermo, Macerata, and Pesaro e Urbino, the Marche Region is home to approximately 1.5 million people and has a population density of approximately 161.6 persons per km^2 in 2020 (Eurostat 2020). The uniqueness of the region lies in its varied topography. In the west, the Apennine Mountains dominate the landscape with their peaks, rolling hills and wide valleys; in the east, the terrain slopes gently down to the Adriatic coast, revealing a coastline characterized by sandy beaches, rocky cliffs, and medium and small towns. The intricate interplay of geographic features has shaped the region's economic activities, such as agriculture, industry, including manufacturing and crafts, and tourism, especially along the coastline. Marche experiences a diverse climate that reflects its varied landscapes. Along the coast, there is a Mediterranean influence that favors mild winters and warm summers. Inland and at higher elevations, the climate tends to be more continental, with colder winters and occasional snowfall in the mountainous areas. Seismic activity is a concern, as the region is classified as "Zone 2", which means that the area can be affected by fairly strong earthquakes (DPC 2023). In fact, the Marche has experienced quite devastating earthquakes, such as the strongest in 1781 (Mw 6.51), and more recently in 1997, which caused 11 casualties (Mw 4.31), and in 2016, which caused 299 casualties (Mw 6.07) (Santoni and Morici 2020; Rovida et al. 2022). In addition, the Marche Region is experiencing an increase in temperature, a decrease in precipitation and snow days, with an indication of a drier climate and an increase in extreme events such as heat waves, implying significant challenges for the hydrological and the geological systems (Marche Region 2022; Sangelantoni et al. 2018). In addition, a rise in sea level along the coasts is predicted, as well as an increase in the frequency and intensity of waves, with consequences for several factors, including the maximum sea level (Marche Region 2022).

The southernmost Italian region is Apulia, which extends between the Adriatic and Ionian Seas. Covering approximately 19,541 km^2, Apulia shares borders with Molise and Campania Regions to the west and Basilicata Region to the southwest. It is divided into six provinces: Bari, Barletta-Andria-Trani, Brindisi, Foggia, Lecce and Taranto. By 2020, the region will be home to nearly 4 million inhabitants, with a population density of about 205.4 people per km^2 (Eurostat 2020). As for Apulia landscapes, in the north, the terrain rises gently to meet the foothills of the Apennines, with rolling hills and fertile plains. Towards the south, the landscape expands into a coastal environment, with sandy beaches, rugged cliffs, and medium -small towns. This varied topography is the basis for Apulia diverse economic activities, including agriculture (especially olive oil), industry (mainly mechanical, automotive, and aerospace), and tourism. The climate of Apulia is influenced by its proximity to the coast: a Mediterranean climate characterized by rather hot and not very rainy summers and winters that are not excessively cold and averagely rainy, with abundant precipitation during the autumn season. Similar to the other regions, Apulia faces seismic concerns due to its location even within the "Zone 1" seismic area, where major earthquakes can occur. In fact, the most devastating earthquake in Apulia occurred in 1646 with at least 5000 victims (Mw 6.72), and more recently in 1948 (Mw 5.55) and 2001 (Mw 6.07) (Ranieri 1949; Rovida et al. 2022). Due to the ongoing climate changes, in recent years, the whole Apulia region has experienced an increase in extreme meteorological events, which have caused, with increasing frequency, phenomena such as floods, landslides, droughts, forest fires, water scarcity, heat waves, and coastal erosion (Apulia Region 2023).

From the Croatian side, Primorje-Gorski Kotar County is located in the northwestern part, embedded between Slovenia in the north, Istria County in the southwest and Lika-Senj County in the southeast. Covering an area of approximately 3588 km^2, it is a region rich in diversity due to its coastal and mountainous landscapes. Primorje-Gorski Kotar consists of various municipalities and cities, including Rijeka, the largest urban center. As of 2020, it boasted a population of 281,945 inhabitants, resulting in a density of about 79.3 persons per km^2 (Eurostat 2020). Geographically, the county transitions from the Adriatic coast, marked by scenic bays and islands, to the interior, dominated by the Gorski Kotar Mountain range. This geographical diversity has significantly shaped the county's economic activities, combining maritime trade, tourism and agriculture with the exploitation of natural resources. Seismic activity is high, primarily in certain areas due to their proximity to fault lines (Herak et al. 2011). The climate of the coastal areas tends toward a Mediterranean climate with mild winters and warm summers, while the mountainous terrain experiences more continental conditions with cooler temperatures and significant precipitation. Over the last 40 years, climate change has affected Primorje-Gorski Kotar County through positive trends in mean annual temperature and negative trends in mean annual precipitation (Meteoblue 2023). In addition, monthly temperature anomalies show a positive deviation from the 30-year climate mean, reflecting an increase in the number of warmer months. The Interreg Italy-Croatia project Joint_SECAP (https://programming14-20.italy-croatia.eu/web/jointsecap) highlighted the challenges posed by climate change to this county, such as damage to the water supply sector due to extended periods of drought, increased interventions related to heat stroke in the health sector, and economic damage to the tourism sector.

The central Croatian study area is Šibenik-Knin County, located in central Dalmatia, surrounded by Zadar County to the northwest, Split-Dalmatia County to the southeast, Bosnia and Herzegovina to the northeast, and the Adriatic Sea to the west. Covering an area of approximately 2984 km^2, this county features a mix of coastal and rugged landscapes. Comprising various towns and villages, including Šibenik, the administrative center, the county had an estimated population of 98,899 inhabitants in 2020, resulting in a density of approximately 33.7 persons per km^2 (Eurostat 2020). The territory extends from the Adriatic coast, with its bays and islands, to the interior, characterized by karst formations and the Dinaric Alps. Economic activities are diverse and include tourism, agriculture and a growing industrial sector. Šibenik-Knin County is also prone to seismic activity, with certain areas falling under high hazard zones (Herak et al. 2011). Climatically, the coastal areas experience Mediterranean conditions with mild winters and hot summers, while the hinterland tends to have a more continental climate, characterized by colder winters and higher precipitation. In 2008, the UNDP Human Development Report for Croatia identified vulnerabilities in the coastal areas of Šibenik-Knin County and outlined a cascade of challenges. These include threats to coastal protection from flooding in historic towns such as Šibenik, and erosion concerns in vulnerable sections of the coastline. Water services are being disrupted by changing precipitation patterns, groundwater scarcity, and reduced freshwater sources, affecting tourism, agriculture, and energy production. Agriculture and fisheries face hurdles such as reduced soil moisture, altered plant life cycles, and difficulties in cultivating marine species, while conservation efforts struggle with invasive species and increased risks of wildfires that alter ecosystems. Tourism faces potential declines due to water scarcity and landscape changes, while broader impacts extend to the insurance and banking industries, changing energy demand and production, and potential impacts on shipping and shipbuilding.

Dubrovnik-Neretva County is located in the southernmost part of Croatia, bordering Bosnia and Herzegovina to the east and Montenegro to the south. Covering an area of approximately 1781 km^2, this county also presents a mixture of coastal and rugged landscapes. It comprises several municipalities and cities, including Dubrovnik, a renowned cultural and historical center. As of 2020, it housed 122,449 inhabitants, resulting in a density of about 69.4 persons per km^2 (Eurostat 2020). Economic activities revolve around tourism, agriculture and trade, heavily dependent on the unique geographical features. The county is prone to seismic activity, with areas falling under high and very high hazard zones (Herak et al. 2011). The climate is Mediterranean along the coast, with mild winters and warm summers, while inland areas have more continental tendencies, with colder winters and higher precipitation. Seismic activity is a concern in certain areas, although it varies throughout the country. Recognized changes in climate patterns include rising air temperatures, changing precipitation patterns, extreme weather events, rising sea levels, and changing river flows. The Interreg Italy-Croatia project Joint_SECAP (https://programming14-20.italy-croatia.eu/web/jointsecap) highlighted the challenges posed by climate change, such as drought for agriculture and water supply, heat waves for health care, temperature increases for the environment and biodiversity, and the development of a register of habitats for plant and animal species.

The RESPONSe project has been proposed to connect and facilitate the constructive exchange between scientists, administrators and citizens, essential for the effectiveness of efforts to adapt to and mitigate climate change. In fact, the overall objective of the project was precisely to provide technical support to local administrations to define and adopt smart governance tools, shared with the entire population and effective as a response to the challenges posed by climate change, studying its effects at a local scale.

To achieve this, three specific objectives were identified. The first objective was to improve local climate knowledge by providing a consistent overview of the current and projected state of the climate system over the project target areas, enhancing the current climate monitoring networks of the target areas, and achieving harmonization and comparability of data across the target areas. The improvement of the monitoring systems, specifically for meteorological data for both Italy and Croatia, and for oceanographic data for Croatia, helped to advance the current knowledge and networks at the local level to meet operational weather and ocean forecasting needs and to support weather event monitoring and future climate change studies. The second objective was to improve the engagement and commitment of local authorities in adaptation planning by raising awareness of the impacts of climate change on the target areas among local authorities, promoting policy frameworks to ensure commitment to adaptation agendas, and facilitating institutional change to operationalize climate adaptation. The aim was setting up of adaptation-friendly policy frameworks targeting pilot areas and taking account of current and future climate change vulnerabilities and territorial risks for each pilot area. The third objective was to promote technical support for the development of local adaptation strategies by supporting local capacity building in the RESPONSe study areas through the provision of personalized technical and decision-making support systems to integrate adaptation measures into existing planning processes.

Specifically, the RESPONSe project activities were divided into four phases. The first phase involved an in-depth analysis of the local climate and monitoring capabilities. A coherent overview of the current and projected state of the climate system in the project study areas was provided. The activities of this phase focused on (a) joint analysis of climate change of observed meteorological and oceanographic fields and time series (2 m air temperature, total precipitation, sea temperature and salinity, and mean sea level) for the target regions, (b) joint analysis of regional atmospheric and ocean climate model (RCM) simulations (both original and bias-adjusted/corrected time series), and (c) introduction of new and/or upgrading of existing meteorological and climatological observing systems for the project focus regions. The results of the analysis consisted of public reports and databases available on the project and partner websites. The climatological analyses were based on state-of-the-art methods, while the regional climate simulations were based on the results of pan-European and pan-Mediterranean projects and initiatives (EURO-CORDEX and Med-CORDEX). Both aspects allow a direct transferability of the project solutions to other European regions, countries, and organizations.

The second phase consisted of a comprehensive analysis of the strategies available to address the critical issues identified in the first phase. This phase is divided into a "top-down" approach and a "bottom-up" approach. The "top-down" approach consisted of identifying the "state of the art" (current conditions) of adaptation and mitigation strategies for the study areas along the Adriatic coast and comparing them with current and future trends in climate variables such as sea level, precipitation and drought. The bottom-up approach consisted in involving stakeholders in the processes of identifying the most relevant adaptation and mitigation strategies needed for their territories. Awareness-raising events, questionnaires, and interviews were used to determine the level of general awareness of the impacts of climate change.

The third phase involved the creation of the Climate Menu for Adriatic Regions (http://www.climatemenu.eu/en/), a toolkit of adaptation actions to initiate adaptation planning at the local level. These actions were grouped and adapted to the characteristics of the three Adriatic macro-regions (northern, central, southern). The Climate Adaptation Menu actions defined objectives, measures, strategies and timelines for the implementation of adaptation planning, taking into account the specific climate change challenges identified in the first phase, the best practices identified in the second phase and the local needs identified in the third phase. The toolkit is available in Croatian, Italian and English and is designed to be easy to understand and therefore easy to use. Public authorities in the Adriatic Basin were able to select from this list those measures that best addressed their local needs with respect to the characteristics of the macro-region to which they belong. This provided the basis for the development of adaptation planning strategies and related decision-making.

The fourth and final phase consisted of the local implementation of the adaptation actions covered in the previous phases. Activities in this phase focused on (i) establishing the framework to promote adaptation planning in the study areas, (ii) conducting vulnerability assessment and risk analysis in relation to context-specific current and future climate change challenges for each study area, (iii) identifying adaptation options adapted to local circumstances, and (iv) providing monitoring and evaluation tools. The Climate Menu of the third phase and the results of the bottom-up approach of the second phase have been used to identify potential adaptation actions for specific Adriatic coastal areas and to accelerate the process of developing effective adaptation plans that address specific needs. The aim is to support local capacity building in the RESPONSe study areas by providing personalized technical and decision support systems to integrate adaptation measures into existing planning processes. The main output of this phase is the development of a Sustainable Energy and Climate Action Plan (SECAP) for each study area. The SECAP, promoted by the European Union's Covenant of Mayors initiative, is a comprehensive strategy developed at the local or regional level to outline specific measures and initiatives to mitigate the effects of climate change. It is a structured framework of policies, actions and targets focused on promoting renewable energy sources, improving energy efficiency, reducing carbon emissions, promoting sustainable transport, improving waste management and encouraging community engagement to achieve environmental sustainability and resilience. A SECAP is expected to achieve the goal of a 40% reduction in greenhouse gas emissions by 2030.

Notably, from the second phase onward, the role of stakeholders has become protagonist. From the assessment of vulnerabilities, intervention priorities, types of actions to be undertaken, to the integration of local knowledge in the analytical processes: in each phase, up to the drafting of the SECAP, the voices and expectations of communities and local authorities have become an integral part of a process that aims to promote a profound, integrated and shared transformation of these coastal realities. Engaging with local administrations is a winding and demanding process that requires time and determination. Nevertheless, this process is fundamental to increase awareness and to ensure the effectiveness of the final results.

In the following chapters, the book extensively describes the second and third phases of the RESPONSe project. First, existing climate change adaptation and mitigation strategies are examined at the European, national and regional levels. The hierarchical structure allows a detailed examination of methodologies, data structures and summaries for each strategy level. The study then compares regional climate trends with these strategies and assesses their effectiveness and gaps. The subsequent chapter is dedicated to paving the way for the implementation of these strategies through an open database of climate adaptation actions. The inclusion of proposed adaptation actions tailored to the study Adriatic region provides practical insights and potentially sets future pathways for research and application in this critical area.

References

Agency for Energy Management of Friuli-Venezia Giulia—ARPA FVG (2023) Segnali dal Clima in FVG. ARPA FVG

Apulia Region (2023) Indirizzi per la stesura della Strategia Regionale di adattamento ai cambiamenti climatici (SRACC). Regione Puglia

Department of Civil Protection—DPC (2023) Seismic classification. https://rischi.protezionecivile.gov.it/en/seismic/activities/emergency-planning-and-damage-scenarios/seismic-classification/

Eurostat (2020) Total population (persons) by regions. https://ec.europa.eu/eurostat/cache/RCI/#?vis=nuts2.population&lang=en

Herak M, Allegretti I, Herak D, Ivančić I, Kuk V, Marić K, Markušić S, Sović I (2011) Republic of Croatia, seismic hazard map. http://seizkarta.gfz.hr

Marche Region (2022) Piano regionale di adattamento al cambiamento climatico (PRACC). Allegato A. Regione Marche

Meteoblue (2023) Climate change Gorski Kotar. https://www.meteoblue.com/en/climate-change/gorski-kotar_croatia_3214579

Ranieri L (1949) Sul periodo sismico dell'estate 1948 in Puglia. Boll Soc Geogr Ital 83(86):213–219

Rovida A, Locati M, Camassi R, Lolli B, Gasperini P, Antonucci A (2022) Catalogo Parametrico dei Terremoti Italiani (CPTI15), versione 4.0. Istituto Nazionale di Geofisica e Vulcanologia (INGV). https://doi.org/10.13127/CPTI/CPTI15.4

Sangelantoni L, Gioia E, Marincioni F (2018) Impact of climate change on landslides frequency: the Esino river basin case study (Central Italy). Nat Hazards 93:849–884. https://doi.org/10.1007/s11069-018-3328-6

Santoni G, Morici R (2020) Terremoti storici nelle Marche costieri—collinari—appenninici e sub-appenninici. Quaderni del Consiglio Regionale delle Marche, (330). Ancona

Slejko D, Cecić I, Riuscetti M (eds) (2019) The 1976 Friuli earthquake: lessons learned. Forum Editore

Chapter 4
Investigating Current Adaptation (and Mitigation) in the Adriatic: The Methodological Foundations

One of the main challenges facing public authorities at different levels of government is the implementation of an effective strategy to deal with the impacts of climate change. For this reason, a rigorous analysis of existing policies can help institutions select the best practices needed to integrate climate change adaptation and mitigation measures into their planning processes. The review of the current strategies is a starting point for planning a response in terms of adaptation and mitigation in a specific area, in order to contextualize the efforts, improve the exchange of experiences, establish an exchange of information, consolidate communication between countries, and improve social learning (Biesbroek et al. 2010).

This work is mainly addressed to public authorities and the scientific community, as it is the only example of a comprehensive data collection on adaptation and mitigation strategies for the Adriatic Basin. Such unique review will serve as a reference for the actions to be taken to implement a mitigation and adaptation strategy in a specific area. In this way, the project represents a first step toward further collaborative research in this emerging field, in order to improve the exchange of experiences, establish a dialogue between countries, and enhance social learning.

4.1 Analysis Criteria

In this work, the development and the current status of mitigation and adaptation strategies in the Italian and Croatian territories surrounding the Adriatic basin have been assessed. In order to achieve this, the analysis has been extended to the European level, to the national level, and to the northern, central, and southern macro-areas of the Adriatic basin. The Regions of Friuli-Venezia Giulia, Marche, and Apulia were selected as representative of the three macro-areas from the Italian side, whereas the Counties of Primorje-Gorski Kotar, Šibenik-Knin, and Dubrovnik-Neretva were selected as representative of the three macro-areas from the Croatian side.

E. Gioia, *Climate Change Adaptation in the Adriatic Coastal Areas*, Disaster Risk Reduction, https://doi.org/10.1007/978-981-97-6655-0_4

The criteria chosen for selecting the strategies to be analyzed are simple: they have to deal with the impacts of climate change, either with solutions aimed at adaptation, mitigation or both.

4.2 Data Structure

In order to standardize the information needed from all the collected strategies and to have a guide during the analysis, a pre-compiled format was realized. The questions of the format were divided into different topics: (i) general information; (ii) context; (iii) relevance; (iv) implementation; (v) interactions; (vi) coherence; (vii) efficiency; (viii) achieved results; (ix) references (see the Annex).

General Information

The general information section aims to provide a foundational understanding of the strategy, encompassing key details like its name, strategy type (mitigation, adaptation, or both), the document framing the strategy (e.g., European Directive, national strategy, regional law), the issuing agency, the year of publication, and the geographical area covered (e.g., Croatia, Marche Region, etc.).

Context

The context section requests a description of the context in which the strategy was developed. In addition to the general objectives and a summary description of the strategy, it asks for the factors that inspired its development. Several motivating factors explain why governments decide to develop and implement a strategy, including compelling legislation or key events that, combined, convince governments and other stakeholders of the need for action. These factors range from the impact of extreme weather events to European policies, insights from United Nations summits and reports, the economic costs of inaction, scientific research, media influence, NGO advocacy, private sector interests, the identification of new opportunities, and learning from examples in other countries or regions.

Relevance

In the relevance section, it is requested to point out the science and policy interactions that emerged in the development and implementation of the strategy, by examining: (a) the scientific basis around the measures, namely the connection with scientific research on climate systems, vulnerability, risks and disasters; (b) the involvement of bridging organizations between science and policy which may either fulfill a coordinating role, provide advice on adaptation policy making, or be responsible for drafting the strategy.

The specific questions posed are as follows:

- Is the strategy based on climate scenarios and projections, considering geographical specificities and best available science?

- Are solid assessments of vulnerabilities/climate risks carried out to support the adaptation decision-making process?
- Does the strategy explicitly take cross-border risks into account, when relevant?
- Does any scientific organization/institution participate in the strategy? Which ones?
- Do any bridging organizations/institutions between science and policy participate in the strategy? Which ones?

Implementation

The implementation section requires an assessment of the "putting into practice" of the strategy, i.e., the sectors involved in the strategy, the time frame, the type and description of actions envisaged, and the preferred policy instruments considered.

The vulnerable sectors and issues involved in the strategy can be identified among the following: agriculture, biodiversity/ecosystem conservation, coastal management, communications, desertification and drought, emergency and rescue services, electricity supply, fisheries, floods and landslides, foreign policy, human health, industry, land use, liability and insurance, tourism and leisure, transport and infrastructure, urban settlement, water resources management, and other (to be specified).

The considered classes of time spans of the strategy, from the beginning to the end of implementation, are "short term" (up to 1 year); "medium term" (1–5 years); "long term" (more than 5 years).

The types of measures identified in the strategy can be based on technical solutions ("gray" measures), based on the improvement of ecosystem services such as tree planting ("green" measures), and non-structural measures focused on individual and societal behavior such as awareness and information, regulation and planning, management, economics, and finance ("soft" measures), or other (to be specified).

The policy instruments foreseen to implement the strategy are: regulatory, which are of a preventive or sanctioning nature, usually combined with some monitoring mechanisms; economic, which work by promoting certain behaviors and practices through economic incentives; voluntary, which aim to promote certain behaviors through voluntary action; communication-related, which aim to provide information about the direction and ambition of government in a particular policy area so that stakeholders can make better-informed choices; other (to be specified) (Bengtsson et al. 2010).

Interactions

The interaction section seeks to identify the possible multi-level interactions in the development and implementation of the strategy by analyzing the stakeholders involved, their roles and responsibilities, and the nature of communication exchanges.

The purpose is to identify which and how public and private stakeholders (e.g., agencies, institutions, citizens) are involved in activities related to the strategy. Moreover, the analysis of the prevailing communication flow allows to recognize either a "top-down" approach, which relies on higher authority figures to set larger goals, or a "bottom-up" approach, which features a decision-making process that gives the entire community a voice in the goals.

Coherence

The coherence section is designed to understand whether the strategy is integrated into new or existing policies, and if so, which ones.

Efficiency

The efficiency section considers the expected timing of implementation and the reason for any deviation. Therefore, it asks for the expected start and end dates of implementation and the status of implementation, whether not started, in progress, delayed or completed. If there is any deviation, the form requires to specify it.

In addition, it is requested to indicate the funding mechanism and the allocated budget (in €).

Achieved Results

The achieved results section aims to analyze the achievement of the overall objectives of the strategy.

For example, it examines whether and how the results of the strategy are evaluated and whether there is a designated body responsible for monitoring the strategy. In addition, the Strengths, Weaknesses, Opportunities, and Threats (SWOT) technique (Weihrich 1982) is used as a tool to evaluate the strategies in order to highlight the main advantages, disadvantages, and lessons learned from their implementation.

References

The references section includes up-to-date links and appendices to the original documents from which the strategy described can be retrieved. The strategy references have been integrated into the chapter References section.

4.3 Data Collection

This data collection was carried out with the help of all the partners involved in the RESPONSe project, coordinated by the Polytechnic University of Marche.

In presenting the adaptation and mitigation strategies within this book, it is crucial to note that the information contained herein stems from data collected between September 2019 and April 2020. As such, while these insights offer a comprehensive review of strategies within the north-central-south Adriatic basin sub-areas, they might not reflect the most recent developments or advancements in this dynamic field. This work strives to provide a foundational understanding rather than an exhaustive update on current practices. Readers are encouraged to supplement this knowledge with the latest research and evolving methodologies in the domain of adaptation and mitigation strategies.

References

Bengtsson M, Hotta Y, Hayashi S, Akenji L (2010) Policy tools for sustainable materials management: applications in Asia. Discussion paper, SCP 2010-001. Institute for Global Environmental Strategies. https://www.jstor.org/stable/pdf/resrep00758.4.pdf

Biesbroek GR, Swart RJ, Carter TR, Cowan C, Henrichs T, Mela H, Morecroft MD, Rey D (2010) Europe adapts to climate change: comparing national adaptation strategies. Glob Environ Change 20(3):440–450. https://doi.org/10.1016/j.gloenvcha.2010.03.005

Weihrich H (1982) The TOWS matrix: a tool for situational analysis. Long Range Plan 15(2):54–66. https://doi.org/10.1016/0024-6301(82)90120-0

Chapter 5
European Framework Strategies for Climate Change

In the context of global climate change, the European Union takes center stage internationally with its endeavors to bolster governance among Member States steering them toward actions essential for effectively tackling this global challenge and shaping a sustainable future. This section dives into European adaptation and mitigation strategies, examining policies and initiatives aimed at reducing emissions, protecting fragile ecosystems, and promoting a more equitable and resilient society.

5.1 Comparing the European Adaptation and Mitigation Strategies

General Information

The European Union (EU) has put in place two key strategies, the EU Adaptation Strategy (EU Adapt) and the EU Emissions Trading System (EU ETS), to tackle the impacts of climate change and reduce greenhouse gas emissions. These strategies, issued by the European Commission, have different objectives, implementation, and outcomes. EU Adapt focuses on adaptation, while EU ETS focuses on mitigation They were issued by the European Commission in 2013 and 2005, respectively, and their geographic coverage includes the then European member states (28 countries) plus Iceland, Liechtenstein, and Norway.

Context

The overall objective of the EU Adapt Strategy is to contribute to a more climate-resilient Europe by enhancing the preparedness and capacity to respond to the impacts of climate change at local, regional, national, and EU levels, developing a coherent approach and improving coordination. The Commission encourages all Member States to adopt comprehensive adaptation strategies (15 had strategies by mid-2013)

E. Gioia, *Climate Change Adaptation in the Adriatic Coastal Areas*, Disaster Risk Reduction, https://doi.org/10.1007/978-981-97-6655-0_5

and will provide guidance and funding to help them build their adaptation capacity and take action. The Commission will also support adaptation in cities by launching a voluntary commitment based on the Covenant of Mayors initiative. The Commission will address knowledge gaps on adaptation and will develop the European Climate Adaptation Platform (Climate-ADAPT) as a "*one-stop-shop*" for adaptation information in Europe. It will also support adaptation in key vulnerable sectors through agriculture, fisheries, and cohesion policy, ensure that Europe's infrastructure is made more resilient and promote the use of insurance against natural and man-made disasters.

The overall objective of the EU ETS strategy is to cost-effectively reduce greenhouse gas emissions by 20% by 2020 and at least 40% by 2030 (compared to 1990 levels). The strategy is based on the "cap and trade" principle. The total amount of greenhouse gases that can be emitted each year by the power plants, factories, and other companies covered by the scheme is subject to a cap set at EU level. Within this Europe-wide cap, companies receive or buy emission allowances, which they can trade if they wish; each allowance gives the holder the right to emit one ton of carbon dioxide (CO_2) or the equivalent amount of two other powerful greenhouse gases, nitrous oxide (N_2O) and perfluorochemicals (PFCs). Allowances can only be used once. Companies must surrender allowances for each ton of CO_2 (or equivalent amount of N_2O or PFCs) covered by the EU ETS that they emitted in the previous year. They face heavy fines if they do not surrender enough allowances to cover their emissions. Companies can get some allowances for free from governments. To cover the rest of their emissions, they must do one or a combination of the following: buy additional allowances; use any surplus allowances they have saved from previous years. Within limits, they can also buy credits from certain types of approved emission-saving projects around the world.

Motivators for both strategies include the implementation of European policies and United Nations summits and reports. EU Adapt is also based on observed and projected climate change impacts, while EU ETS is aligned with the Kyoto Protocol's greenhouse gas reduction targets. EU ETS is also influenced by private sector interests and the recognition of new opportunities.

Relevance

In terms of relevance, the strategies are developed on the basis of climate scenarios and projections, taking into account specific geographical factors and the best available scientific information. They also consider the greenhouse gas reduction targets set out in the Kyoto Protocol. For example, EU Adapt is based on observed and projected climate change and impacts for the main biogeographical regions in Europe (EEA 2012). On the other hand, the EU ETS is based on the greenhouse gas reduction targets reported under the Kyoto Protocol.

EU Adapt prioritizes EU-wide vulnerability assessments to support decision-making processes, considering both natural and man-made risks across different sectors, and recognizes transboundary climate risks, i.e., risks which cross administrative boundaries. It highlights how climate risks operate across borders due to the myriad of complex and global linkages between people, ecosystems and economies (e.g., trade, international financial flows, migration, and security). Conversely, the EU ETS lacks vulnerability assessments and does not consider transboundary risks to be materially relevant. However, it should be noted that transboundary risks are not relevant in this strategy, since one of the main objectives is to involve many companies in the carbon market.

In terms of scientific organizations and institutions involved in this strategy, the Joint Research Centre is actively involved in EU Adapt, while EU ETS does not directly involve scientific entities.

In terms of bridging organizations that facilitate collaboration between science and policy, the Copernicus Climate Change Service (C3S) is involved in EU Adapt, while several climate and energy related institutions are involved in EU ETS. These include the Directorate-General for Climate Action (DG CLIMA), the European Energy Exchange (EEX), the common auction platform, and the Intercontinental Exchange (ICE) as the UK auction platform.

Implementation

Both strategies cover a wide range of sectors and issues, some of which overlap, such as industry, and responsibility and insurance. EU Adapt also covers agriculture, communications, and fisheries, while EU ETS covers transport and infrastructure. Both are expected to have a duration of more than 5 years from start to end of implementation, using a mix of measures including "green" and "soft" typologies (even "gray" for EU Adapt). Basically, EU Adapt: Encourage all Member States to adopt comprehensive adaptation strategies; Provide LIFE funding to support capacity building and strengthen adaptation action in Europe (2014–2020); Include adaptation in the Covenant of Mayors (2013/2014); Close the knowledge gap; further develop Climate-ADAPT as a one-stop-shop for adaptation information in Europe; facilitate the climate-proofing of the Common Agricultural Policy (CAP), the Common Fisheries Policy (CFP); ensure more resilient infrastructure; promote insurance and other financial products for resilient investment and business decisions. On the other hand, EU ETS: Establish a carbon price and assign a financial value to each ton of emissions saved in the carbon trading market. Impose an EU-wide cap on emissions, reduce them by 1.74% annually, and gradually move to auctioning allowances rather than allocating them for free. Encourage heavy industry, manufacturing, and aviation to invest in clean technologies and low-carbon solutions.

To implement these measures, both strategies envisage the use of economic and financial policy instruments. The EU Adapt Plan will also use voluntary and communication instruments.

Interactions

Stakeholders, either public or private (e.g., agencies, institutions, citizens), are involved in the development and implementation of both strategies.

In EU Adapt, the main stakeholders are Member States; regional and local authorities and decision-makers; financial institutions; private companies and industries in the energy, transport, and buildings sectors; civil society; conservation practitioners; insurers. Member States are encouraged to develop their own adaptation strategies, which will be assessed by the European Commission, to identify gaps in knowledge on adaptation and the relevant tools and methodologies to address them. Regional and local authorities and decision-makers are encouraged to adopt local adaptation strategies (e.g., Covenant of Mayors), integrate them into the Common Agricultural Policy (CAP), Cohesion Policy and the Common Fisheries Policy (CFP), and organize awareness-raising activities. Member States and regions can also use Cohesion Policy and CAP funding 2014–2020 to address knowledge gaps, invest in the necessary analyses, risk assessments and tools, and build capacities for adaptation. Public authorities and decision-makers, civil society, the private sector, and conservation practitioners need to ensure the full mobilization of ecosystem-based approaches to adaptation. Relevant industries need to work on infrastructure and physical assets to climate-proof vulnerable investments. Insurers are encouraged to unlock the full potential of insurance pricing and other financial products for risk prevention, mitigation and long-term resilience in investment and business decisions.

The main stakeholders in the EU ETS are electricity and heat producers; energy-intensive industries, including oil refineries, steel mills and the production of iron, aluminum, metals, cement, lime, glass, ceramics, pulp, paper, cardboard, acids, and organic bulk chemicals; civil aviation. Their responsibility is to comply with the "rules" of the carbon market and thus to surrender used allowances; companies operating in the covered sectors must participate in the EU ETS as it is mandatory.

The communication approach in EU Adapt is a mixture of "top-down" and "bottom-up" methods, whereas the EU ETS is strictly "top-down".

Coherence

EU Adapt is integrated with various sectoral policies, such as: the LIFE 2014–2020 environmental funding program; the Common Agricultural Policy; the Cohesion Policy; the Common Fisheries Policy; the Green Paper on insurance against natural and man-made disasters. In contrast, the EU ETS lacks specific integration.

Efficiency

Both strategies have specific implementation timelines, with EU ETS still in progress. The initial end dates for implementation are 2013–2020 for EU Adapt and 2005–2030 and beyond for EU ETS. There are no reported deviations for either strategy. The European Commission adopted its new EU Climate Change Adaptation Strategy 2021–2027 on February 24, 2021. The new strategy sets out how the European Union will adapt to the unavoidable impacts of climate change and become climate resilient by 2050.

The budget allocated to EU Adapt is financed by the European Union, while the EU ETS does not have a dedicated budget. The EU institutions have agreed to allocate at least 20% of the EU budget for 2014–2020 to climate adaptation, which is three times the previous level. During this period, the EU allocated €220.8 billion to climate action, or 20.59% of the total budget, exceeding the 20% target. The EU's long-term budget for 2021–2027, together with the NextGenerationEU recovery instrument, amounts to €2.018 trillion in current prices (or 30% of the EU budget). It will be used to repair the economic and social damage caused by the coronavirus pandemic and to support the transition to a modern and more sustainable Europe (European Commission 2023a).

On the other hand, the total amount of the fund for EU ETS depends on the carbon price. According to the European Commission, "following the increase in the carbon price to about €80 per ton of CO_2 in 2022 (from €55 in 2021), Member States revenues from the Emissions Trading System (ETS) have doubled in the course of two years to almost €30 billion in 2022. Prices are expected to remain well above €55 per ton of CO_2 in the years to come" (European Commission 2023b).

Achieved Results

Both strategies have been evaluated by the European Commission and the Union registry, respectively, for EU Adapt and EU ETS. In the first case, the Commission published an evaluation of the strategy in November 2018. The evaluation was accompanied by a public consultation from December 2017 to March 2018 (page available in most EU languages). The analysis resulted in a report on lessons learned and reflections on improvements for future action, accompanied by a staff working document presenting the evaluation in detail. In the second case, the accurate accounting of all allowances issued is ensured by a single Union registry with strong security measures. The registry keeps track of the ownership of allowances held in electronic accounts, just as a bank keeps track of its customers and their money.

The EU Adapt Strategy has met its objectives, with progress recorded for each of its eight individual actions. The main strengths have been the cooperation between different policy areas and platforms, and the involvement of local authorities and communities in the Member States' discussions, in line with the bottom-up approach. However, the actions of the strategy could be better integrated. Links between actions were spontaneous but could be better identified and exploited. Similarly, the strategy should better integrate the international dimension of adaptation In order to synchronize with global collective policies and actions on sustainable development, biodiversity and disaster risk reduction, among others. The strategy promoted adaptation plans at all levels but was less effective in implementing and monitoring these plans in Member States. Progress on the number of local adaptation strategies has been slower than envisaged in 2013 and varies between States.

The main strength of the EU ETS was that the need to use up their reserves of allowances and credits created a permanent incentive for companies to reduce emissions by investing in more efficient technology or switching to less carbon-intensive energy sources; companies could also sell allowances and credits, for example, if they judged that they had more than they needed: This allowed them to choose the most cost-effective options for managing their emissions. However, there are challenges associated with allowances and credits. In the first trading period, the number of allowances based on estimated needs proved to be too high; in the second trading period, the number of allowances was reduced by 6%. In the second trading period, the number of allowances was reduced by 6.5%, but the economic downturn reduced emissions and thus demand even more: this led to a surplus of unused allowances and credits that continues to weigh on the carbon price; "carbon leakage" refers to the situation that can occur when companies, because of the costs associated with climate policies, relocate production to other countries that have less stringent restrictions on greenhouse gas emissions: this could lead to an increase in their overall emissions.

The main lessons learned highlight the importance of EU Adapt in preparing for climate impacts and the success of the EU ETS in creating a carbon market.

The EU Adapt Strategy has been a reference point for preparing Europe at all levels for the impacts of climate change. The Commission will continue to work toward its objectives. The EU adaptation strategy remains highly relevant and adds value to national, regional, and local adaptation efforts while remaining cost-effective. The strategy, a policy instrument with low administrative burden for most stakeholders, has succeeded in focusing decision-makers on the need to prepare for climate risks. The strategy has acted as a reference point, focusing and catalyzing action at other levels of government. It has also successfully channeled efforts to ensure that climate change considerations continue to be integrated into policies and budgets at the EU level.

The EU ETS has put a price on carbon and proven that trading greenhouse gas emissions works. The carbon market creates an incentive to invest in technologies that reduce emissions and drives clean investment in developing countries. The first trading period successfully established free trading of allowances across the EU, put in place the necessary infrastructure and developed a vibrant carbon market. The environmental benefits of the first phase may be limited by the over-allocation of allowances in some Member States and some sectors, mainly due to the reliance on emissions projections before verified emissions data became available under the EU ETS. When the publication of verified emissions data for 2005 revealed this "over-allocation", the market reacted as expected by lowering the market price of allowances. The availability of verified emissions data has allowed the Commission to ensure that the cap on national allocations in the second phase is set at a level that leads to real emission reductions. In addition to underscoring the need for verified data, experience to date has shown that greater harmonization within the EU ETS is essential to ensure that the EU achieves its emission reduction targets at least cost and with minimal distortion of competition. The need for greater harmonization is most evident in the way the overall cap on emissions allowances is set. The first two trading periods also show that widely differing national methods for allocating allowances to installations threaten fair competition in the internal market. Further harmonization, clarification and refinement are also needed in relation to the scope of the scheme, access to credits from emission reduction projects outside the EU, the conditions for linking the EU ETS to emissions trading schemes elsewhere, and monitoring, verification, and reporting requirements.

5.2 Summary

In response to the UN summit, the European Commission has implemented two different strategies based on climate scenarios: EU Adapt for adaptation (2013) to climate change and EU ETS for mitigation (2005) to reduce greenhouse gas emissions. The focus of the first strategy is a resilient Europe, the implementation of a guideline and the establishment of the Covenant of Mayors, while the goal of the second strategy is to set percentage values to be achieved for the reduction of greenhouse gas (GHG) emissions, 20% by 2020 and 40% by 2030, such as CO_2, N_2O, PFCs, and heavy metals. In the adaptation strategy, the Commission plans to fill the knowledge gaps on adaptation and to develop the European Climate Adaptation Platform (Climate-ADAPT) as a "*one-stop shop*" for information on adaptation in Europe. In the mitigation strategy, companies receive or buy allowances that they can trade if they wish; each allowance gives the holder the right to emit one ton of CO_2 or the equivalent amount of two other potent greenhouse gases, N_2O and PFCs.

Both strategies include vulnerable sectors and issues such as industry and responsibility and insurance, and the time span is long term (over 5 years). The adaptation strategy also includes the primary sector, such as agriculture and fisheries, and the tertiary sector, while the mitigation strategy also includes transportation and infrastructure. The measures envisaged are "green" and "soft" for both strategies, and "gray" for adaptation. The policy instruments envisaged to implement the strategy are economic/financial instruments for both strategies, and voluntary and communication-related for adaptation. Top-down communication is predominant for both strategies, but the adaptation strategy also uses bottom-up communication. Their implementation is ongoing, the funds are European and at the end they will be evaluated with public consultation and by the EU Commission and the Union Registry. The main strengths are the cooperation and the reduction of polluting gases to reduce costs and carbon price, but there are weaknesses like the slower progress for the local communities and the carbon leakage.

References

European Commission (2005) EU emissions trading system (EU ETS). https://ec.europa.eu/clima/policies/ets_en#tab-0-1

European Commission (2013) EU adaptation strategy. https://climate.ec.europa.eu/eu-action/adaptation-climate-change/eu-adaptation-strategy_en

European Commission (2023a) Supporting climate action through the EU budget. https://climate.ec.europa.eu/eu-action/eu-funding-climate-action/supporting-climate-action-through-eu-budget_en#eu-budget-2014-2020-climate-integrated-into-all-spending-areas

European Commission (2023b) EU budget: commission puts forward an adjusted package for the next generation of own resources. https://ec.europa.eu/commission/presscorner/detail/en/ip_23_3328
European Environment Agency—EEA (2012) Climate change, impacts and vulnerability in Europe 2012. EEA report no. 12/2012. Publications Office of the European Union, Luxembourg. https://www.eea.europa.eu/publications/climate-impacts-and-vulnerability-2012

Chapter 6
Climate Strategies at National Level: Comparing Italian and Croatian Approaches

The 28 European countries (plus Iceland, Liechtenstein and Norway) must translate the EU adaptation and mitigation strategies into national strategies. They use the guidelines and adapt the strategy to the natural and social characteristics of their territories. This section examines the characteristics of the Italian and Croatian national strategies.

6.1 Italy

General Information

Italy has put in place three key strategies to address the impacts of climate change and reduce greenhouse gas emissions: the National Adaptation Strategy (Italian Ministry for the Environment and the Protection of Land and Sea 2015; hereafter "NAS"), the National Plan for Adaptation to Climate Change (Italian Ministry for the Environment and the Protection of Land and Sea, n.d.; hereafter "NPACC"), and the Integrated National Energy and Climate Plan (Italian Ministry of Economic Development, Ministry of the Environment and of the Land and Sea Protection, and the Ministry of Infrastructure and Transport 2018; hereafter "INECP").

The NAS and the NPACC are both strategies developed for Italy by the Italian Ministry for the Environment and the Protection of Land and Sea (MATTM). The NAS, published in 2015, and the NPACC, committed in 2016, focus primarily on addressing the impacts of climate change within the country. Both fall under the category of adaptation strategies, which aim to address the challenges posed by climate change.

On the other hand, the INECP, released in 2018, takes a different approach. It is a mitigation strategy formulated by several ministries—the Ministry of Economic Development, Ministry of the Environment and of the Land and Sea Protection, and the Ministry of Infrastructure and Transport. The INECP focuses on integrating

E. Gioia, *Climate Change Adaptation in the Adriatic Coastal Areas*, Disaster Risk Reduction, https://doi.org/10.1007/978-981-97-6655-0_6

energy and climate planning to mitigate the effects of climate change, an approach that differs from the NAS and NPACC.

All three strategies—NAS, NPACC, and INECP—have been specifically designed to address climate-related issues in Italy and underscore the nation's commitment to addressing climate challenges through a combination of adaptation and mitigation strategies, albeit with different emphases and issuing authorities.

Context

The overall objective of the NAS is to provide a framework for adapting to the consequences of climate change and to lay the foundations for: improving the current knowledge on climate change and its impacts; describing the vulnerability of the territory and the adaptation options; promoting participation; supporting awareness and information on adaptation to climate change; specifying the tools to be used. The strategy first considers the expected impacts of climate change and the main vulnerabilities in Europe and the Mediterranean; then it focuses on the expected impacts of climate change and the main vulnerabilities in Italy, referring to the sectors of socio-economic and environmental importance that are most vulnerable to climate change. Finally, it assesses the state of knowledge on the impacts and vulnerabilities of these sectors and possible intersectoral aspects.

The general objective of the NPACC is to offer a support tool to national, regional, and local institutions for the identification and selection of the most effective actions in the different climatic areas in relation to the criticalities that characterize them most and for the integration of adaptation criteria in existing procedures and tools. This objective is divided into four specific purposes: (i) to limit the vulnerability of natural, social, and economic systems to the impacts of climate change, (ii) to increase their adaptive capacity, (iii) to improve the exploitation of any opportunities and (iv) to facilitate the coordination of actions at different levels.

The general objective of the INEPC is to align with the European energy and environmental objectives and the reduction of greenhouse gas emissions. Specifically, the diffusion of renewable energy and energy efficiency, related to the reduction of polluting and climate-changing emissions, the improvement of energy security and increasing economic and employment opportunities for families and the production system. The approach aims to put citizens, including prosumers, and businesses, in particular small and medium-sized enterprises, increasingly at the center. At the same time, it will be necessary to assess the compatibility of energy and climate objectives with the objectives of protecting the landscape, air quality, water, biodiversity, and soil.

The motivating factors behind the development of these strategies include different elements. All of them draw motivation from extreme weather events/impacts and the implementation of European policies and United Nations summits/reports. In addition, the adaptation strategies also include scientific research behind the need to adapt to climate change.

Relevance

National strategies rely heavily on climate scenarios and projections and, in the case of NAS, also incorporate reports from prominent scientific bodies and numerous studies on economic impacts (McCallum et al. 2013; Carraro 2008). These analyses take into account geographical specificities and are supported by a comprehensive assessment of vulnerabilities and climate risks. This assessment in the NAS covers a wide range of potential impacts, including temperature increases, heat waves, changes in precipitation, drought, sea level rise, impacts on human health and biodiversity, water resources, and socio-economic impacts. In the NPACC, a synthetic risk index has been calculated for the Italian territory by integrating the overall index of potential impacts (derived from the combination of hazard indicators with those of exposure and sensitivity) and that of adaptive capacity. In addition, in the INECP, the effort to assess vulnerability extends to the evaluation of the most vulnerable population groups, such as the elderly and children.

Moreover, while the NAS does not explicitly address transboundary risks, the other strategies recognize the importance of recognizing such risks and the need for cooperation among states or coordination with other countries and organizations, especially in the INECP, due to intraday market dynamics and cross-border balancing markets.

Several scientific organizations and institutions have contributed significantly to the adaptation strategies. In the NAS, the Euro-Mediterranean Center on Climate Change (CMCC) coordinated a technical table of about one hundred national experts from universities, research institutes and foundations. Notably, in the NPACC, a wide range of institutions are actively involved, such as the National System for Environmental Protection (SNPA), the National Research Council (CNR), the Institute of Biometeorology (IBIMET), the Institute of Sciences of the Atmosphere and Climate (ISAC), the National Agency for New Technologies, Energy and Sustainable Economic Development (ENEA), and various universities, including the Ca' Foscari University of Venice, the University of Bologna, the Polytechnic University of Marche, among others.

However, in terms of bridging organizations or institutions between science and policy, there seems to be limited or no explicit involvement.

Implementation

The three national strategies cover a range of sectors and vulnerabilities, with a focus on specific areas. The NAS includes a wide range of vulnerable sectors, including agriculture, biodiversity/ecosystem conservation, coastal management, desertification and drought, electricity, fisheries, floods and landslides, human health, industry, liability and insurance, tourism and leisure, transportation and infrastructure, urban settlements, and water resources management. In addition, the NAS highlights forests and mountainous areas as particularly vulnerable. Notably, the NPACC covers the full range of vulnerable sectors considered here. The INECP, on the other hand, outlines vulnerability in selected sectors, namely agriculture, human health, industry, land use, transport and infrastructure, as well as in the residential, waste, tertiary, and forest sectors.

In terms of timeframe, all strategies envisage "long-term" implementation, focusing on actions to be carried out over more than 5 years.

In terms of measures, the adaptation strategies (NAS and NPACC) include a mix of "gray", "green", and "soft" measures, while the mitigation strategy (INECP) includes only "gray" measures.

Basically, the actions in the NAS are divided into two types: those to be implemented by 2020 and those to be implemented after 2020, and for each time period they have specified the "gray", "green" and "soft" actions to be undertaken for the action sectors. In general, the short-term action concerns the planning, evaluation, optimization management of resources, actions and planning of economic instruments to manage climate risk. The long-term actions involve the definition of specific indices of climate change variables, the adoption of new codes and measures related to the different sectors, the adoption of the most sophisticated information and technology systems and the implementation of awareness campaigns.

In the NPACC, the "soft" measures are the following: to plan the information strategy through research and evaluation, monitoring, dissemination, awareness and training; organizational and participatory processes to select the genotypes and genetic varieties, organize the civil protection at the local level, diversify the business strategies, controls and inspections; creation of tables, committees, networks; at the level of governance, legislative and regulatory adjustments, plans and strategies, and economic and financial instruments must be implemented; finally, guidelines are necessary. On the other hand, the "green" and "gray" measures are: adaptation and improvement actions of facilities and infrastructures, planning of reforestation and afforestation; solutions based on ecosystem services, planning of conservation, reconstruction and renaturalization of coastal areas, taking into account urban development.

Finally, the INECP focus on the following: increasing the diffusion of low-emission and renewable technologies; promoting technological innovation to develop new tools with high potential; improving energy efficiency by containing system costs; phasing out coal-fired power plants by 2025 in a fully secure manner; ensuring the security and flexibility of gas and electricity networks; closing the gap on electricity prices for households and businesses; reducing the risk of relocation of more energy-intensive companies. In particular, strategic environmental assessment procedures (VAS) are foreseen in each sector. The reduction of emissions in the waste sector is mainly linked to the increase of separate waste collection and subsequent recycling. In the energy sector, Italy has planned the gradual phase-out of coal-fired electricity generation by 2025, with a first significant step in 2023, offset by the rapid growth of renewable energies and a plan for infrastructural interventions to be carried out in the coming years. With regard to measures in the agricultural and livestock sector, the Regions are required to adopt practices aimed at reducing emissions from agricultural activities, such as the covering of manure storage facilities, the use of correct methods of spreading manure and the burying of land to be fertilized, where such practices are technically feasible and economically sustainable. In the forestry sector, the aim is to know the Italian forests in a timely manner and to train operators to recognize the state of abandonment of the forest and the forest management and planning guidelines are mandatory. All sectors are expected to reduce some atmospheric pollutants such as sulfur dioxide, nitrogen oxides, non-methane volatile organic compounds, ammonia, and particulate matter.

To implement these measures, all strategies envisage the use of regulatory (except NAS) and communication tools.

Interactions

Stakeholders, either public or private (e.g., agencies, institutions, citizens), are involved in the development and implementation of all strategies.

In the NAS, the main stakeholders are the Ministry of the Environment and Protection of the Land and Sea (MATTM), the Ministry of Agriculture and Forestry Policies, the Ministry of Infrastructure and Transport, the Ministry of Health, the Ministry of Cultural and Environmental Heritage, the Ministry of Education, University and Research, the Ministry of Economic Development, the Ministry of Regional Policy, Tourism and Sport, the Civil Protection, the Committee of the Regions, the National Association of Italian Municipalities, the Union of Italian Provinces and the European Mediterranean Center for Climate Change (CMCC). Representatives of each of the Ministries involved participated in the Institutional Table convened by the MATTM. The MATTM is the body responsible for climate policy at the national level and has managed the definition of the NAS to be implemented through sectoral action plans. The CMCC has managed the technical-scientific coordination to obtain the information necessary for the development of the NAS.

In the NPACC, the main actor is the MATTM, which implements the strategy based on the NAS.

On the other hand, in the INECP, the main stakeholders are the Ministry of Economic Development, the MATTM and the Ministry of Infrastructure and Transport, which were responsible for drafting the plan to adapt it to European energy and climate regulations.

The communication approach is mainly "top-down", but also "bottom-up" in the NAS.

Coherence

The NAS was adopted by a Decree of the Director General of Climate and Energy in June 2015, and thus integrated into sectoral policies. This decree committed the Director General to develop the NPACC in order to implement the NAS. The INECP was adopted by a resolution of the Council of Ministers on April 9, 2019.

Efficiency

All strategies have specific timelines for implementation, and all are ongoing.

The initial end dates for implementation are 2015–2020 and beyond for NAS, 2016–2050 for NPACC, and 2018–2050 for INECP. There are no reported deviations for either strategy.

The NAS refers to European funds that need to be established in the NPACC. Some actions have already been taken and the funding comes from specific projects, such as: €372.000 for the use of naturalistic engineering techniques to improve the functionality of the River Serio in Lombardy; €80 million Horizon 2020 program for scientific research; €198 million INTERREG Alcotra (Italy-France) for planning strategy; €97 million Maritime INTERREG (Italy-France) for the protection and evaluation of the environmental heritage; €30 million INTERREG Italy-Switzerland) for tools for cross-border water resources; 24 million INTERREG Italy-Austria for connecting and enhancing habitats; €39 million INTERREG Italy-Slovenia for promoting energy efficiency strategies and action plans; €188 million INTERREG Italy-Croatia for marine innovations; €81 million INTERREG Italy-Greece for safeguarding marine and terrestrial biodiversity; €23 million INTERREG Italy-Malta for safeguarding marine and terrestrial biodiversity; €139 million INTERREG AlpineSpace for sustainable mobility actions; 163 M INTERREG CentralEurope for energy efficiency of public infrastructure; €60 M INTERREG Adrion for sustainable protection and enhancement of cultural and natural heritage; €125.80 INTERREG Med for energy efficiency of public buildings; €109 M ENI-CBC MED for measures to save and sustainably manage water resources; €92 M IPA-CBC Italy-Albania-Montenegro for measures to safeguard biodiversity and reduce environmental risks; 33 million INTERREG Italy-Tunisia for environmental protection; €19.7 million European Investment Bank for transport infrastructure; €500 million for the 2015–2020 National Action Plan; €140 million program for the methanization of southern Italy; €90 million for the reclamation and conversion of sites of national interest; €1 billion "Culture and Tourism" level; €60 million for the rehabilitation of school buildings. And many others reported in the PNACC.

INECP also refers to European funds, and €180 billion euros have been allocated for the national energy system between 2017 and 2030.

Achieved Results

Evaluation of results has been implemented for all strategies. The evaluation mainly consists of monitoring, with different levels of detail.

In terms of oversight, for NAS, the responsibility for monitoring and reviewing the strategy lies with the institutional administrations, indicating a shared governance approach. For NPACC, the MATTM at the national level and the Regional Environmental Protection Agencies (ARPAs) at the regional and local levels are the designated bodies for monitoring and review. For INECP it is the MATTM.

The main strengths identified in these strategies include the NAS as the reference point for coordinated adaptation actions in the country and the provision of a common guideline in the NPACC. On the other hand, weaknesses include the lack of coordination between the authorities involved and the wide range of sectors covered.

Interestingly, no specific lessons learned or emerging needs/opportunities are explicitly outlined in the information provided, leaving this aspect unaddressed or possibly unexplored in the strategy documents.

6.2 Croatia

General Information

Croatia has adopted two key strategies to address the impacts of climate change and reduce greenhouse gas emissions: the Climate Change Adaptation Strategy in the Republic of Croatia for the period to 2040 with a view to 2070 (Croatian Ministry of Environmental Protection and Energy 2020; hereafter "CAS"), and the Energy Development Strategy of the Republic of Croatia until 2030 with a view to 2050 (Croatian Ministry of Environmental Protection and Energy 2021; hereafter "LCDS").

These strategies include different approaches to address climate change in Croatia. The CAS focuses primarily on adaptation measures and was officially adopted as a national strategy by the Croatian Parliament on April 7, 2020. In contrast, the LCDS emphasizes mitigation efforts and aims to transition Croatia to a low-carbon economy, having been officially adopted as a national strategy in June 2021.

Both strategies, issued by the Ministry of Environment and Energy, focus on Croatia as a geographical area and represent the country's commitment to address climate change through both adaptation and mitigation measures, demonstrating a comprehensive approach to climate action.

Context

The overall objective of the CAS is to promote a plan that will make Croatia resilient to climate change. The long-term goal is to reduce the vulnerability of natural systems and society to the negative impacts of climate change, i.e., to increase the resilience and ability to recover from these impacts. The strategy aims to describe the Croatian way to prepare all important segments of the economy and society for climate changes. The document analyzes motivation for development of strategy, goals and priorities in plans for climate adaptation. The important part of the strategy are adaptation measures to climate change (with description and stakeholders), implementation plan and overall environmental impact of the strategy.

The general objectives of the LCDS are as follows: achieving sustainable development based on a low-carbon, resource-efficient, knowledge based and competitive economy; increasing the security and sustainability of energy supply, increasing the availability of energy and reducing energy dependency; solidarity by fulfilling Croatia's obligations under international agreements, within the framework of European Union policies, as part of our historical responsibility and contributing to global goals; reducing air pollution. The document provides the legal framework for the development of the strategy and explains the current situation in Croatia. The strategy defines emission targets for periods up to 2020, 2030 and 2050 and provides modeled values for emissions. Low carbon scenarios are defined for each sector (transport, agriculture, industry ...). The strategy also considers the financial segment of the projects (source of financing), environmental, social, and economic impacts of the predicted scenarios.

The motivating factors behind the development of these strategies include different elements. All are induced by scientific research, examples from other countries, and the implementation of European policies and United Nations summits/reports. In addition, the mitigation strategy also includes the impact of extreme weather events behind the need to address climate change.

Relevance

The Croatian strategies rely heavily on climate scenarios and projections. In the case of CAS, the strategy is based on the most relevant climate projection models with projections for the period 2011–2040 and 2041–2070. The model uses available data for the period 1971–2000 for all key impact indicators (sea level, rainy/dry days …). The strategy is the result of an integral approach, which, starting from the analysis of the situation in selected sectors and cross-sectoral thematic areas, determines determinants, cross-sectoral impacts and vulnerabilities, as well as aggregate measures, taking into account the possibilities of plan implementation and the cross-sectoral impacts of implemented measures. In the case of the LCDS, the strategy is based on climate scenarios and projections for each sector in the defined time frame. Greenhouse gas emissions and projections are reported according to the methodology of the Intergovernmental Panel on Climate Change (IPCC 2006) and the corresponding greenhouse potential factors for the calculation of CO_2 emission equivalents. LCDS provides an analysis of climate risks and the impacts of climate change on the economy, society and the environment. However, neither explicitly addresses transboundary risks.

Several scientific organizations and institutions were invited to provide input or comments on the preparation of the document.

Implementation

The CAS and LCDS in Croatia address different sectors and vulnerabilities. Both focus on agriculture, fisheries, tourism and leisure, urban settlement, water resource management and energy. In addition, the CAS focuses on biodiversity and human health. On the other hand, LCDS also prioritizes sectors such as electricity, industry, land use, transport and infrastructure, and waste management.

The CAS focuses on "long-term" actions, while the LCDS envisages actions over "short-term" and "long-term" time horizons. The "long-term" measures extend their vision to 2070 and 2050, respectively. Both strategies emphasize a mix of "gray", "green", and "soft" measures. All measures are defined with timelines, descriptions, investments, and expected results.

Both strategies include regulatory and economic/financial instruments for implementation, although LCDS also includes communication-related measures.

Interactions

Government and related ministries, as well as scientific institutions, are involved in the development of both Croatian strategies, and, through consultations, public debates, in the implementation of the expected measures.

The communication approach is mainly "top-down".

Coherence

The CAS was integrated into the United Nations Framework Convention on Climate Change (UNFCCC), the Climate-ADAPT; the UN Sustainable Development Goals (SDGs), the Kyoto protocol, and the Paris Agreement on Climate Change. The LCDS was integrated into the Kyoto protocol, the Paris Agreement on Climate Change, and the Roadmap for moving to a competitive low carbon economy in 2050 (European Commission 2011).

Efficiency

All strategies have specific timelines for implementation. The initial end dates for implementation are 2020–2070 for CAS and 2020–2050 for LCDS. Implementation is at an early stage and there are no reported deviations for either strategy.

The CAS refers to European funds such as European Regional Development Fund (ERDF), Cohesion Fund, European Maritime and Fisheries Fund (EMFF), European Agricultural Fund for Rural Development (EAFRD), European Social Fund (ESF). Currently, 3.6 million euros have been allocated.

LCDS also refers to European funds such as European Structural and Investment Funds (ESI), EU Emissions Trading System (EU ETS) funds and emission taxes. The current allocated budget is €50 million until 2030.

Achieved Results

Evaluation of results is foreseen for all strategies. Evaluation will mainly be done by monitoring specific indicators.

Responsibility for monitoring and evaluation lies with different government agencies. The CAS involves ministries and local units, while the LCDS identifies agencies, ministries and local units that are responsible for specific target results.

Identified strengths of the CAS include the technical and financial identification and assessment of measurable targets, which provide a clear framework for assessing progress. However, weaknesses were not explicitly identified. In the LCDS, the strengths are not explained, while the concerns are about the surplus of unused allowances and credits in the carbon market due to overestimation. Indeed, in the first trading period, the number of allowances based on estimated demand proved to be excessive; in the second trading period, the number of allowances was reduced by 6.5%, but the economic downturn depressed emissions and thus demand even more: In addition, considerations of "carbon leakage" are highlighted, addressing the potential shift of production to countries with less stringent emissions regulations, which could lead to an increase in their total emissions.

The lessons learned from CAS underscore the need for immediate action by all relevant stakeholders to prepare society for the climate change challenges ahead. However, in the case of LCDS, no specific lesson or emerging need/opportunity is explicitly mentioned, leaving this aspect unaddressed or possibly unexplored in the strategy.

6.3 Summary

The analysis collated data pertaining to five national adaptation and mitigation strategies in Italy and Croatia.

Italy has formulated three pivotal strategies to combat climate change and curb greenhouse gas emissions: the National Adaptation Strategy (NAS), the National Plan for Adaptation to Climate Change (NPACC), and the Integrated National Energy and Climate Plan (INECP). Published between 2015 and 2018, these strategies differ in their approaches. The NAS and NPACC, developed by the Italian Ministry for the Environment, primarily focus on addressing climate impacts within the country, while the INECP, formulated by multiple ministries, integrates energy and climate planning to mitigate climate effects.

These national strategies heavily rely on climate forecasts and projections, incorporating reports from significant scientific bodies and economic impact studies in the NAS. These assessments consider geographic specifics, evaluating vulnerabilities and climate risks comprehensively. The NAS evaluates a wide array of potential impacts like temperature shifts, heat waves, precipitation changes, drought, rising sea levels, and their effects on health, biodiversity, water resources, and socioeconomics. The NPACC crafted a synthetic risk index for Italy, merging potential impact indices with adaptive capacity assessments. Although the NAS does not explicitly address cross-border risks, the NPACC and INECP acknowledge their significance, emphasizing cooperation among states and organizations.

Diverse scientific entities contribute substantially to these strategies. The NAS engages the Euro-Mediterranean Center on Climate Change, orchestrating a technical forum involving numerous national experts from academia and research foundations. Similarly, the NPACC integrates various entities such as research centers and universities. However, there's limited explicit involvement of intermediary organizations or institutions bridging science and policy in these strategies.

All Italian initiatives operate on a long-term trajectory exceeding 5 years, with all strategies primarily relying on "gray" measures. However, the NAS and NPACC also incorporate "green" and "soft" measures. The NAS outlines two distinct types of measures: immediate actions to be executed by 2020 and a framework defining climate change indices and future measures. Conversely, the NPACC focuses on information strategy, financial instrument implementation, and infrastructure planning, while the INECP proposes renewable technologies and environmental assessment procedures. Policy instruments supporting these strategies are primarily communication-oriented across the board, with regulatory tools featured in the NPACC and INECP. Communication in these strategies generally follows a "top-down" approach, although the NAS integrates a "bottom-up" communication segment.

All strategies have specific timelines for implementation, that commonly end in 2050, and all are ongoing. These strategies integrate seamlessly into existing or new sector policies. Funding is facilitated by the EU, and the strategies' outcomes are assessed through monitoring mechanisms.

The main strength of the NAS is the realization of a strategic document and for the NPACC Strategy the efficacy point is the common guideline that is free consulting. The poor coordination of the authorities is considered as the main weakness of the NPACC.

Croatia has adopted two central strategies to combat climate change: the Climate Change Adaptation Strategy (CAS) aiming to fortify Croatia against climate impacts until 2040, and the Low-Carbon Development Strategy (LCDS) striving for a low-carbon economy until 2030. The CAS primarily focuses on adaptation, while the LCDS emphasizes mitigation efforts, both symbolizing Croatia's comprehensive approach to climate action, formulated and endorsed by the Ministry of Environment and Energy.

CAS and LCDS share a focus on vital sectors like agriculture, fisheries, tourism, urban areas, water resources, and energy. However, the CAS uniquely targets biodiversity and human health. Conversely, LCDS places priority on a broader spectrum, including electricity, industry, land use, transportation, infrastructure, and waste management, aiming for a comprehensive approach to low-carbon development across various sectors.

Both strategies rely on climate scenarios and projections, considering scientific research and European policies. The CAS integrates diverse measures and funding from European sources, while the LCDS bases projections on IPCC methodology. Although both strategies involve government bodies and scientific institutions, communication mainly follows a "top-down" approach.

These strategies focus on long-term actions, with an emphasis on a mix of regulatory, economic, and communication-related measures. Integrated into international frameworks, the CAS aligns with the UNFCCC and the Paris Agreement, whereas the LCDS connects to the Kyoto Protocol and the European Commission's roadmap for a low-carbon economy.

Both strategies are in early implementation stages without reported deviations, with allocated funds from various European sources. However, the evaluation of results is planned mainly through monitoring. Strengths in the CAS include measurable targets for progress assessment, while the LCDS faces challenges with unused allowances in the carbon market, economic downturns impacting emission reductions, and concerns about potential carbon leakage.

The CAS emphasizes the necessity for immediate action by all stakeholders to confront climate challenges. However, the LCDS lacks explicit mention of lessons learned or emerging needs, leaving these aspects unexplored within the strategy.

In summary, the strategies analyzed at the national level reflect the indications of the EU adaptation and mitigation strategies to implement specific strategies for their territories. In fact, the characteristics between the Italian and the Croatian national strategies are similar. The objective of the Italian and Croatian adaptation strategies is to prepare a resilient nation to climate change. The mitigation Strategies focus on achieving the reduction of greenhouse gas emissions by promoting renewable energy and energy efficiency. All national strategies include almost all vulnerable sectors, and the common approach is top-down. The Italian mitigation strategies address transboundary risks. All national measures have a "long-term" time horizon.

All Italian and Croatian strategies include "gray", "green" and "soft" measures. The funding comes from the EU and monitoring is widespread to evaluate the planned activities. The strategies of both nations have the intention to create a document with the main objectives for a good adaptation and mitigation planning.

References

Carraro C (2008) Cambiamenti climatici e strategie di adattamento in Italia. Una valutazione economica. Il Mulino, Bologna

Croatian Ministry of Environmental Protection and Energy (2020) Climate change adaptation strategy in the Republic of Croatia for the period until 2040 with a view to 2070. https://mingor.gov.hr/UserDocsImages/KLIMA/Climate%20change%20adaptation%20strategy.pdf

Croatian Ministry of Environmental Protection and Energy (2021) Low carbon development strategy of the Republic of Croatia until 2030 with a view to 2050. https://mingor.gov.hr/UserDocsImages/klimatske_aktivnosti/odrzivi_razvoj/NUS/lts_nus_eng.pdf

European Commission (2011) Roadmap for moving to a competitive low-carbon economy in 2050. COM(2011) 112 final. https://eur-lex.europa.eu/LexUriServ/LexUriServ.do?uri=COM:2011:0112:FIN:EN:PDF

Intergovernmental Panel on Climate Change—IPCC (2006) 2006 IPCC guidelines for national greenhouse gas inventories. In: Eggleston HS, Buendia L, Miwa K, Ngara T, Tanabe K (eds) Prepared by the national greenhouse gas inventories programme. IGES, Japan

Italian Ministry for the Environment and the Protection of Land and Sea (2015) Strategia Nazionale di Adattamento ai Cambiamenti Climatici [National adaptation strategy]. https://www.minambiente.it/sites/default/files/archivio/allegati/clima/strategia_adattamentoCC.pdf

Italian Ministry for the Environment and the Protection of Land and Sea (n.d.) Piano Nazionale di Adattamento ai Cambiamenti Climatici [National plan for adaptation to climate change]. https://www.mase.gov.it/pagina/piano-nazionale-di-adattamento-ai-cambiamenti-climatici

Italian Ministry of Economic Development, Ministry of the Environment and of the Land and Sea Protection, and the Ministry of Infrastructure and Transport (2018) Proposta di Piano Nazionale Integrato per l'Energia e il Clima Italiano [Integrated national energy and climate plan]. https://www.mise.gov.it/images/stories/documenti/Proposta_di_Piano_Nazionale_Integrato_per_Energia_e_il_Clima_Italiano.pdf

McCallum S, Dworak T, Prutsch A, Kent N, Mysiak J, Bosello F, Klostermann J, Dlugolecki A, Williams E, König M, Leitner M, Miller K, Harley M, Smithers R, Berglund M, Glas N, Romanovska L, van de Sandt K, Bachschmidt R, Völler S, Horrocks L (2013) Support to the development of the EU strategy for adaptation to climate change: background report to the impact assessment, part I—problem definition, policy context and assessment of policy options. Environment Agency Austria, Vienna

Chapter 7
Regional Climate Strategies in the Adriatic: Translating European and National Agendas

Regional adaptation and mitigation strategies are an integral part of broader national and European climate change policies. Cooperation, alignment, and coordination between these levels are essential to effectively address climate change and to ensure a comprehensive approach that addresses both local challenges and global climate goals.

This section examines the characteristics of the Italian and Croatian regional strategies, divided into northern, central, and southern Adriatic strategies.

7.1 Northern Adriatic Strategies

This section examines the current mitigation and adaptation strategies implemented in the northern Adriatic area, specifically focusing on the Friuli-Venezia Giulia Region for the Italian side and the Primorje-Gorski Kotar County for the Croatian side.

7.1.1 Friuli-Venezia Giulia Region

General Information

The Friuli-Venezia Giulia Region has implemented a diverse array of strategies addressing climate change adaptation and mitigation across various sectors (Table 7.1):

(1) Burden Sharing Friuli-Venezia Giulia (Italian Ministry of Economic Development, 2012; hereafter "Burden Sharing");

E. Gioia, *Climate Change Adaptation in the Adriatic Coastal Areas*, Disaster Risk Reduction, https://doi.org/10.1007/978-981-97-6655-0_7

Table 7.1 List of strategies issued in the Friuli-Venezia Giulia Region, including the indication of the type of strategy and the year of issue

Strategy	Mitigation	Adaptation	Year
Burden sharing	X		2012
Climate impacts		X	2015
PAI		X	2017
Energy plan	X		2015
PSR 2014–2020	X	X	2015
Air quality	X	X	2007
Cognitive study		X	2018
Transport infrastructure	X		2011
Budoia Charter		X	2017
NOEMIX	X		2017
Green purchases	X		2018
TRUST		X	2009
Environmental criticalities	X	X	2013

(2) Impacts of climate change on the regional physical territory (Friuli-Venezia Giulia Autonomous Region, 2015a; hereafter "Climate Impacts");
(3) Plan for the Geo-hydrological Management (Friuli-Venezia Giulia Autonomous Region, 2017; hereafter "PAI");
(4) Regional Energy Plan (Friuli-Venezia Giulia Autonomous Region, 2015b; hereafter "Energy Plan");
(5) Rural development program 2014–2020 (Friuli-Venezia Giulia Autonomous Region, 2015c; hereafter "PSR 2014–2020");
(6) Regional plan for the improvement of the air quality (Friuli-Venezia Giulia Autonomous Region, 2007; hereafter "Air Quality");
(7) Cognitive study of climate change and some of their impact in Friuli-Venezia Giulia (Friuli-Venezia Giulia Agency for Environmental Protection, 2018; hereafter "Cognitive Study");
(8) Regional plan of transport infrastructure, freight mobility and logistic (Friuli-Venezia Giulia Autonomous Region, 2011; hereafter "Transport Infrastructure");
(9) Budoia Charter for the Alps (Alpenallianz, 2017; hereafter "Budoia Charter");
(10) NOEMIX—New Mobility in Friuli-Venezia Giulia (European Union Horizon 2020 NOEMIX, n.d.; hereafter "NOEMIX");
(11) Friuli-Venezia Giulia Region action plan for green purchases (Friuli-Venezia Giulia Autonomous Region, 2018; hereafter "Green Purchases");
(12) Tool for regional scale assessment of groundwater storage improvement in adaptation to climate change (European Commission LIFE TRUST, n.d.; hereafter "TRUST");
(13) Main Environmental criticalities and response actions for the territory of the Friuli-Venezia Giulia Region (Friuli-Venezia Giulia Autonomous Region, 2013; hereafter "Environmental Criticalities").

These initiatives cover a wide range of areas, such as burden sharing, climate change impacts on the regional territory, geo-hydrological planning, energy, rural development, air quality improvement, cognitive studies on climate change impacts, transportation infrastructure, green purchasing, groundwater storage assessment, and environmental criticality.

The strategies vary in their emphasis on adaptation and mitigation efforts. While some focus solely on adaptation or mitigation, others have a dual objective that combines both (Table 7.1). These strategies were issued in the years 2007–2018, focusing primarily on the Friuli-Venezia Giulia Region itself, with some extending to Alpine (Budoia Charter) or neighboring (TRUST) areas.

Most of the strategies have been adopted through regional plans of the Friuli-Venezia Giulia Autonomous Region, while a few others have been adopted through decrees of the Italian Ministry of Economic Development (Burden Sharing), reports of the Regional Environmental Agencies, i.e., Regional Environmental Protection Agencies, and EU projects (Climate Impacts, Cognitive Study, NOEMIX, TRUST) or the voluntary declaration of the Alpine Alliance (Budoia Charter).

Context

The general objective of the Burden Sharing is to promote the use of renewable energy and reduce energy consumption. A regional target has been set to achieve a 17% share of renewable energy in Italy's gross final energy consumption (electricity, heat and transport) by 2020. This strategy is in line with the European 20–20–20 target of reducing greenhouse gas emissions by 20% compared to 1990 levels, increasing the share of renewable energy to 20% and improving energy efficiency by 20%.

The overall objective of the Climate Impacts strategy is to analyze the impacts of climate change with a deep knowledge of the regional climate and terrain (impacts on surface and underground water systems, coastal areas, and geo diversity). The document foresees a framework on the possible impacts of climate change on the geomorphologic and geo-hydrological system in Friuli-Venezia Giulia and the definition of an adaptation strategy, with scientific basis and monitoring of the effects, for a correct regional and local planning.

The main objective of the PAI is the mitigation of hydraulic and geological hazards in the river basins of regional competence (i.e., Slizza, Levante, Lagoon of Marano and Grado). This plan works in the field of land management, with particular reference to the protection of the population and human settlements at risk. The document promotes maintenance actions against landslides and floods to increase the stability and hydraulic conditions of the territory.

The overall purpose of the Energy Plan is to increase the rational use of energy with smart, sustainable and inclusive growth. The plan identifies objectives and measures for the development and improvement of the regional energy system in relation to the global vision of reducing greenhouse gas emissions and increasing energy efficiency (20–20–20 goal). Fifty-seven actions have been identified for the region.

Two are the main objectives of the PSR 2014–2020: (i) to contribute to the improvement of the competitiveness of agriculture and primary producers and (ii) to contribute to the preservation and enhancement of ecosystems. The objectives are broken down into six priorities: promote knowledge transfer and innovation in agriculture and forestry in rural zones; strengthen the profitability of farms and the competitiveness of agriculture in all its forms in all regions and promote innovative technology for farms and sustainable forest management; promote the organization of the food chain, including the processing and marketing of agricultural products, animal welfare and risk management; preserving, restoring, and enhancing ecosystems related to agriculture and forestry; promoting the efficient use of resources and the transition to a low-carbon and climate-resilient economy in the agri-food and forestry sectors; working toward social inclusion, poverty reduction and economic development in rural areas.

The Air Quality strategy aims to define scenarios for consolidating and protecting air quality in order to achieve pollution levels that respect the limits set by current legislation. The objectives mainly concern the reduction of ozone and particulate matter.

The overall objective of the Cognitive Study is to support the preparation of a regional strategy for adaptation and mitigation of climate change. National climate change impacts will be used as a basis for the regional analysis.

The main purpose of Transport Infrastructure is to create synergies between services and modes of transport. The establishment of an efficient transport system is considered as one of the fundamental elements to achieve the objective of creating a competitive market and territorial cohesion. A strategic environmental assessment (VAS) is used to guarantee the sustainability of the choice of plan and to integrate environmental considerations in order to minimize the impact and propose the appropriate mitigation and compensation measures.

The aim of the Budoia Charter is to create an instrument for implementing the principles of the Alpine Convention (https://www.alpconv.org/en/) at the local level in order to implement a series of public policies for sustainable development in the Alps. The Alpine municipalities commit themselves to implement local measures for adaptation to climate change, to undertake actions to assess the potential risks and opportunities for the municipal territory, to promote public debate and to raise awareness among citizens, residents and visitors of the risks and opportunities related to climate change at the local level. This is done to ensure the safety of the territory in the face of the expected impacts of climate change in the Alpine Space at all levels of administration, and especially at the level closest to the individual living space.

NOEMIX is a mitigation strategy for the transition to a low carbon economy. The objectives are the reduction of CO_2 emissions, the increase of energy production by renewable energy, the reduction of pollution, the reduction of specific matter and noise generated by congestion in urban areas. The strategy consists of a car-sharing service for public administration through the substitution of obsolete vehicles. The Friuli-Venezia Giulia Region intends to play a key role in the transition through electric mobility in synergy with the Energy Plan and SECAP of the municipalities involved.

The overall goal of Green Purchasing is to transition to more sustainable production and consumption practices and to support sustainable growth. The strategy aims to encourage the public administration to choose goods and services with low environmental impact in the fields of construction, energy supply, transport and IT equipment, in order to have a significant impact on the reduction of greenhouse gas emissions.

The aim of the TRUST strategy is to provide climate change hazard scenarios for the shallow groundwater of the Veneto and Friuli-Venezia Giulia plains in northern Italy. The document identifies adaptation measures based on artificial aquifer recharge to mitigate the effects of drought and water scarcity. It also assesses the potential impacts of climate change on surface and groundwater and related ecosystems, in order to provide appropriate information for the sustainable management of water resources and the definition of effective adaptation options.

Finally, the main objective of the Environmental Criticalities strategy is to promote an environment conducive to business innovation and efficient management of natural resources. The goal is to identify mitigation and adaptation strategies related to the primary environmental criticalities.

The motivating factors behind the development of these strategies include various elements. Most of them are driven by the impact of extreme weather events and the implementation of European policies. The Climate Impacts and TRUST adaptation strategies are also based on scientific research, while the Energy Plan, Transport Infrastructure and Green Purchases mitigation strategies are based on United Nations summits/reports. Finally, the Energy Plan also recognizes the possibility of new opportunities and NOEMIX includes examples from other countries/regions behind the way to address climate change.

Relevance

The Friuli-Venezia Giulia strategies rely heavily on climate scenarios and projections. For example, Climate Impacts uses meteorological networks to monitor regional impacts and forecast models for decision-making, while PAI focuses on the study of significant parts of river basins in the Friuli-Venezia Giulia Region. In addition, Energy Plan provides scenarios of future system evolution (economic, environmental) for the territory and TRUST uses downscaled climate projections of precipitation, temperature and evapotranspiration fields for risk and impact assessment. Environmental Criticalities also includes climate scenarios as one of the thematic areas considered for analysis. Finally, Transportation Infrastructure includes an environmental assessment that analyzes regional environmental contexts.

These analyses generally take into account geographic specificities and are supported by a comprehensive assessment of vulnerabilities and climate risks. This is the case for PAI, PSR 2014–2020, Air Quality, Cognitive Study and Environmental Criticalities. In addition, the socio-economic potential of the measures of the Energy Plan was assessed by analyzing three studies at the national level (Greenpeace, Enel Foundation, Energy Services Manager) and one at the international level (International Renewable Energy Agency). Moreover, in the area of transportation infrastructure, an environmental assessment was carried out through the identification of possible negative effects of the Plan's actions on the environment and the identification of mitigation measures for these effects. In addition, in the Budoia Charter, the Alpine Municipalities undertake to carry out actions aimed at assessing the potential risks and opportunities of climate change for the municipal territory. Furthermore, the Green Purchasing Strategy takes into account the scarcity of natural resources and the excessive production of waste. Finally, in TRUST, three reference scenarios were selected and used to run cascading regional geo-morphoclimatic and geo-hydrological models. The models provided information on the potential variations of water balance components in relation to each climate change scenario. Exceptions to these practices are the Burden Sharing and NOEMIX strategies.

Only the Energy Plan, PSR 2014–2020, Air Quality, Cognitive Study and Budoia Charter strategies explicitly consider transboundary impacts and possible consultation with neighboring regions or states. In particular, in the Budoia Charter, environmental impacts in the surrounding states were evaluated to assess the possibility of activating transboundary consultation with Austria and Slovenia.

Several scientific organizations and institutions were invited to provide input or comments on the development of the strategy, such as the Friuli-Venezia Giulia Agency for Environmental Protection (ARPA-FVG), the Universities of Udine and Trieste, the International Center for Theoretical Physics (ICTP), the National Institute of Oceanography and Geophysics, and the National Research Council—Institute for Biological Resources and Marine Biotechnologies (CNR IRBIM).

The participation of organizations linking science and policy has been reported for the Energy Plan, the PSR 2014–2020, the Air Quality, the Cognitive Study, the Transport Infrastructure and the NOEMIX.

Implementation

The 13 regional strategies cover a broad range of sectors and vulnerabilities, including agriculture, biodiversity/ecosystem conservation, coastal management, communications, desertification and drought, electricity supply, fishery, floods and landslides, human health, industry, responsibility and insurance, tourism and leisure, transport and infrastructure, urban settlements, and water resources management (Table 7.2). Interestingly, communications, and responsibility and insurance are addressed by only one strategy (Air Quality and PAI, respectively), whereas emergency and rescue services, and foreign policies are not addressed at all.

In terms of timeframe, most of the strategies envisage "long-term" implementation, focusing on actions to be carried out over 5 years, with the exception of the Energy Plan, NOEMIX, Green Purchases and TRUST, which envisage "medium-term" implementation. On the other hand, the Budoia Charter does not explicitly mention a specific time span.

In terms of measures, most strategies include only "gray" measures, apart from Air Quality, which includes both "gray" and "soft" measures. Similarly, PSR 2014–2020 includes both "green" and "soft" measures, while Green Purchases includes only "soft" measures. Finally, Environmental Criticality includes all types of measures.

Specifically, the Burden Sharing actions include public transport measures, building improvements, urban traffic reduction, incentives for energy savings in public lighting, and water management. The Climate Impact strategy includes improving irrigation and drinking water networks, expanding the monitoring network, implementing early warning systems, modifying planning policies,

Table 7.2 List of the vulnerable sectors and the Friuli-Venezia Giulia Region strategies that address them

Sector	Strategy engagement
Agriculture	Energy plan, PSR 2014–2020, cognitive study, green purchases, TRUST
Biodiversity/ecosystem conservation	Climate impacts, PSR 2014–2020, cognitive study, green purchases, TRUST, environmental criticalities
Coastal management	Climate impacts, cognitive study, environmental criticalities
Communications	Air quality
Desertification and drought	Cognitive study, TRUST
Electricity supply	Burden sharing, energy plan, PSR 2014–2020, cognitive study, green purchases
Emergency and rescue services	–
Fishery	Cognitive study, environmental criticalities
Floods and landslides	Climate impacts, PAI, PSR 2014–2020, cognitive study, TRUST, environmental criticalities
Foreign policy	–
Human health	Cognitive study, environmental criticalities
Industry	Energy plan, cognitive study
Land use	PAI, cognitive study, NOEMIX
Responsibility and insurance	PAI
Tourism and leisure	Cognitive study, environmental criticalities
Transport and infrastructure	Burden sharing, energy plan, air quality, transport infrastructure, NOEMIX, environmental criticalities
Urban settlement	Energy plan, cognitive study
Water resource management	Burden sharing, TRUST, environmental criticalities

reducing groundwater extraction, managing geo-hydrological instability, monitoring coastal zones, and preserving geodiversity. Mitigating hydraulic and geological hazards in PAI involves various structural (such as the construction of expansion boxes) and non-structural (such as the use of rockfall barriers, wire mesh, and drainage) measures to address these risks. The Energy Plan focuses on transforming traditional energy systems into sustainable ones, increasing energy efficiency across sectors, promoting sustainable energy awareness, developing sustainable mobility, reducing greenhouse gas emissions, and seeking sustainable mechanisms for cross-border infrastructure. The PSR 2014–2020 contains a detailed list covering various aspects of agricultural development, investment in fixed assets, rural renewal, forest development, establishment of producer associations and organizations, climate payments, organic farming, environmental payments, support for areas facing constraints, animal welfare, cooperation and local development strategies. Air Quality includes measures to reduce emissions in the transport and energy sectors through sustainable mobility, incentives for fleet renewal, car and bike sharing, restrictions on fuel consumption, parking strategies, diversification of energy sources, and communication plans for these measures. The Cognitive Study includes preliminary studies on bioclimatic classification, geo-hydrological instability, marine ecosystems, agricultural impacts, forest production, and the impact of climate change on energy consumption. The Transportation Infrastructure strategy aims to rationalize transportation infrastructure, reduce road congestion, develop quality road networks, improve road safety, and encourage the transfer of goods and people to more sustainable modes of transportation such as rail and water. The Budoia Charter calls for efforts to adapt to climate change and increase resilience, especially in the Alpine countries. The NOEMIX is a project that involves replacing a certain number of traditional cars with electric ones, installing charging stations, and ensuring that part of the electricity demand comes from renewable energy. Green Purchases focuses on implementing environmental criteria for purchasing goods and services, integrating social and environmental criteria, reducing energy consumption, and promoting sustainable practices. TRUST discusses the use of modeling tools to assess the impacts of climate change on groundwater and ecosystems, guiding decisions on crop selection, groundwater protection, well development rights, and expected outcomes such as energy savings and reduced CO_2 emissions. Finally, Environmental Criticalities includes actions to improve energy efficiency, use renewable energy, manage ecosystems (agricultural and forest), promote sustainable mobility, monitor and improve resilience in mountainous areas, and manage natural disasters, focusing on biodiversity and physical territory resilience.

To implement these measures, most of the strategies envisage the use of either regulatory or communication policy instruments, or both. Burden Sharing, Energy Plan and NOEMIX also use economic/financial policy instruments for their implementation. Exceptions are the Budoia Charter and TRUST, which only foresee voluntary mechanisms, and the Cognitive Study, which is characterized by exploratory studies only.

Interactions

The development and implementation of the strategies involve a wide range of public and private stakeholders from different sectors.

All Italian regions are involved in the Burden Sharing initiative to reach the percentage target for the reduction of greenhouse gas emissions. In other strategies, such as Climate Impacts, PSR 2014–2020, Cognitive Study, Transport Infrastructure, NOEMIX, Green Purchases and Environmental Criticalities, the Friuli-Venezia Giulia Region plays the key roles of coordination, analysis, collaboration with other entities and institutions, monitoring, and evaluation.

Critical stakeholders include agencies such as Civil Protection, Geological Service and institutions such as the National Agency for New Technologies, Energy and Sustainable Economic Development (ENEA), the Council of Local Autonomies (CAL) and bodies involved in Strategic Environmental Assessment (VAS). They are invited as technical and scientific participants. Involvement extends to a wide range of entities, including regional park authorities, local government bodies, gender policy bodies, human rights committees, and volunteer committees. Representatives from various sectors such as agriculture, forestry, trade unions, non-agricultural economic categories, environmental and consumer associations, academia, and anti-discrimination organizations are integral contributors.

In particular, the involvement of citizens and social sectors is central in the strategies of PSR 2014–2020, Air Quality and Budoia Charter, in order to consider different points of view of all members of the community, to achieve the objectives and to create links with the local institution.

The communication approach is mainly "top-down", except for the strategies characterized by explorative studies or proposals (Climate Impacts, Cognitive Study, and NOEMIX).

Coherence

The strategies are deeply integrated into numerous existing sectoral policies and frameworks at various levels—international, European, national, regional and even local initiatives.

Some strategies, such as Climate Impacts, Energy Plan, PSR 2014–2020, Air Quality, NOEMIX and TRUST, synchronize with global efforts to mitigate and adapt to climate change, strengthening their international compatibility and cooperation. On the other hand, Cognitive Study integrate into the National Adaptation Strategy (NAS) and the National Climate Change Action Plan (PNACC) to ensure consistency and cooperation in addressing climate change impacts at the national level. Furthermore, the integration of transport infrastructure involves regional and local levels, with general transport and logistics plans, urban and regional plans ensuring alignment with transport policy, logistics, air quality improvement, and water protection strategies.

The remaining strategies incorporate European, national and regional policies. For example, Burden Sharing is aligned with the 20–20–20 targets set by the European Commission in 2011 (Energy Roadmap 2050), incorporating national renewable energy consumption targets (Ministerial Decree 15-3-2012) and local government initiatives for sustainable development and energy transition (Covenant of Mayors). PAI is also firmly rooted in established European directives (such as Dir. 2000/60/CE, Dir. 2007/60/CE), national laws (Legislative Decree 152/2006) and regional legislation (Law 16/2002). In the Budoia Charter, the recognition and incorporation of documents and policies adopted at both European and Alpine levels ensures synchronization with broader regional initiatives for climate change adaptation and sustainability. Moreover, the Green Purchasing Strategy is integrated with the European Commission and the Ministry of the Environment Recommendation and with the National Action Plan for Green Purchasing. Finally, the Environmental Criticalities strategy dovetails with overarching European 20–20–20 targets and national laws to ensure consistency in goals and actions toward sustainability and environmental protection.

Efficiency

All strategies have specific timelines for implementation and are either underway or completed, with no reported deviations.

The initial implementation dates range from 2006 to 2018, while the completion dates range from 2011 to 2100, but mostly 2020.

In a few cases, the funding is specified (EU, national, regional) and the allocated budget is reported. For example, in the Energy Plan, the budget is €10.3 million (regional financing); and, in the PSR 2014–2020, the budget is €296 million (€128 million from EU balance, €168 million from national and regional co-financing).

Achieved Results

Evaluation of results has been implemented for most of the strategies. The evaluation methods vary and include different approaches such as monitoring, simulation, and specific assessments with indicators. Monitoring bodies include regional, local, and environmental authorities as well as stakeholders at different administrative levels. Responsibility for monitoring and review lies with several bodies. These include regional bodies responsible for proposing actions if targets aren't met (Burden Sharing, PAI, PSR 2014–2020, Air Quality, Cognitive Study, Environmental Criticalities), environmental agencies such as ARPA FVG (Climate Impacts, Energy Plan, Air Quality, Transport Infrastructure, Environmental Criticalities), and a collaborative network involving regional, provincial and local authorities as well as municipalities within the Friuli-Venezia Giulia region (Air Quality). Stakeholders play a crucial role in monitoring the process, while public procurement observers and stakeholders are involved in specific areas such as procurement related aspects.

Key strengths identified in these strategies include the scientific basis for Climate Impacts and the Cognitive Study. In addition, the Energy Plan and Environmental Criticalities maintain consistency with European legislation and guidelines, prioritizing the protection of human health. In addition, the PSR 2014–2020 strategy capitalizes on Friuli-Venezia Giulia's predominantly rural landscape, robust economic factors, including good income levels and employment rates, and comparatively high educational participation. It also capitalizes on the region's biodiversity and implements effective agricultural practices that reduce net emissions. It also optimizes traffic and tourism flows while maintaining high data quality standards. Innovative modeling techniques provide numerical advantages in understanding pollutant transformations, with cost-effective bio-monitoring methods and inventive tools such as transplantation enhancing their effectiveness. NOEMIX, which has the potential to foster collaboration among small communities, creates a shared sense of concern and a unified effort toward solutions. In Green Purchases, it is underscored by a commitment to quality assurance that emphasizes objective criteria and direct corporate involvement. Finally, the TRUST strategy demonstrates the value of modeling and risk assessment tools for sustainable groundwater management, a model that can be replicated beyond Italy.

On the other hand, the weaknesses identified include the unavailability of data (Climate Impacts and TRUST), some specific features of the region such as the decline of agro-forestry activities and loss of landscape (PSR 2014–2020) or significant hydraulic risk (Cognitive Study), the difficulties in finding resources (Budoia Charter) or carrying out the analyses (Environmental Criticalities).

Interestingly, only in three strategies are specific lessons learned or emerging needs/opportunities explicitly outlined. These are the facts that the creation of a community can produce positive effects (Budoia Charter), that the transition to the use of electric cars is possible for citizens and businesses (NOEMIX), and that the consideration of environmental, energy and social performance and not only the most economically advantageous can be considered (Green Purchases).

7.1.2 Primorje-Gorski Kotar County

General Information

The Primorje-Gorski Kotar County has implemented one climate change mitigation strategy named Program for air protection, ozone layer protection, climate change mitigation, and adaptation to climate change in Primorje-Gorski Kotar County for the period 2019—2022 (Primorje-Gorski Kotar County, 2019; hereafter "PGZ"). The strategy has been issued in the year 2019 by the Primorje-Gorski Kotar County itself.

Context

The general objective of the PGZ program is to define and elaborate objectives and measures affecting air quality (energy, transport, industry, waste management, etc.). The program defines the sequence for implementing these measures, setting deadlines for execution, identifying responsible entities for implementation, and evaluating the program's execution based on prioritized measures and activities. The program consists of the following parts: legal framework for the development, scope, description of the current air quality and emissions, report on the implemented measures from previous plans, goals for the protection of the air and ozone layer and measures for the mitigation of climate change.

The motivating factors behind the development of this strategy include the implementation of European policies and scientific research.

Relevance

The PGZ strategy is based on climate scenarios and projections, using geographic specificities and the best available science to shape its approach. It conducts comprehensive vulnerability assessments and, where relevant, explicitly considers transboundary risks that cross administrative boundaries. However, scientific organizations or institutions are not directly involved in the strategy. Rather, it involves bridging organizations such as fishermen's associations and others focused on environmental issues.

Implementation

The PGZ strategy covers a small range of topics, namely industry and transport and infrastructure, while the other sectors are not addressed at all.

In terms of timeframe, the program foresees a "medium-term" implementation, focusing on actions to be carried out in less than 5 years, namely from 2019 to 2022.

The strategy includes only "soft" measures. Specifically, preventive measures to maintain air quality; short-term measures when there is a risk that the alert threshold will be exceeded; measures to achieve the limit values for certain pollutants in the air within the given deadline if they are exceeded; measures to achieve the long-term objectives for ground-level ozone in the air; measures to reduce and limit emissions of pollutants that cause the adverse effects of acidification, eutrophication and photochemical pollution; measures to reduce and/or limit emissions of heavy metals; measures to phase out controlled substances that deplete the ozone layer; measures to reduce emissions of fluorinated greenhouse gases; measures to reduce and/or limit emissions of greenhouse gases and to adapt to climate change; measures to promote increased energy efficiency and the use of renewable energy; and measures to reduce overall emissions from transport.

The use of either regulatory or voluntary policy instruments is envisaged to implement these measures.

Interactions

The PGZ does not specify if the development and implementation of the strategies involved public and private stakeholders. The communication approach is mainly "top-down".

Coherence

The strategy was created as a legal obligation under the Croatian Air Protection Act (2011), which is the national law that "determines the competence and responsibility for the protection of air and the ozone layer, mitigation of climate change and adaptation to climate change, planning documents, monitoring and assessment of air quality, measures to prevent and reduce air pollution, reporting on air quality and data exchange, quality monitoring activities air and air emissions, substances that damage the ozone layer and fluorinated greenhouse gases, monitoring of greenhouse gas emissions and measures to mitigate and adapt to climate change, air protection information system, financing of air protection, ozone layer, mitigation of climate change and adaptation to climate change, administrative and inspection supervision".

Efficiency

The timeframe of the PGZ was between 2019 and 2022, so the strategy is completed with no reported deviations.

The funding is a co-financed from the European Union and the funds of the Marišćina County Waste Management Center, but the allocated budget is not specified.

Achieved Results

The evaluation of the results is foreseen through the descriptive definition of the actions, with named stakeholders and priority of implementation. The evaluation will be carried out by a designated person responsible for each area covered by the program.

The main strengths of the PGZ were identified as: the grouping of measures for the protection of the air, the ozone layer and the mitigation of climate change, and the adoption of environmental principles, among others, based on respect for generally accepted principles of environmental protection, respect for the principles of international environmental law and respect for scientific knowledge. On the other hand, the main weakness of the program is that it does not provide budgetary information.

7.2 Central Adriatic Strategies

This section examines the current mitigation and adaptation strategies implemented in the central Adriatic area, specifically focusing on the Marche Region for the Italian side and on the Šibenik-Knin County for the Croatian side.

7.2.1 Marche Region

General Information

The Marche Region has implemented a variety of strategies addressing climate change adaptation and mitigation across different sectors (Table 7.3):

Table 7.3 List of strategies issued in the Marche Region, including the indication of the type of strategy and the year of issue

Strategy	Mitigation	Adaptation	Year
GIZC plan	X	X	2019
PEAR 2020	X		2016
Sustainability	X	X	2007
PRGR	X	X	2015
PSR 2014–2022	X	X	2015
Development		X	2017
PFR	X	X	2005
Forest fires 2017–2019	X		2017
Forest Table	X	X	2005
Aqueducts	X		2014

(1) Integrated Coastal Zone Management Plan (Primorje-Gorski Kotar County, 2019; hereafter "GIZC Plan")
(2) Regional Environmental Energy Plan (Marche Region, 2016; hereafter "PEAR 2020")
(3) Regional Environmental Action Strategy for Sustainability 2006–2010 (Marche Region, 2007; hereafter "Sustainability")
(4) Regional Waste Management Plan (Marche Region, 2015a; hereafter "PRGR")
(5) Rural Development Program 2014–2020 (Marche Region, 2015b; hereafter "PSR 2014–2022")
(6) Extraordinary Plan for the development, promotion and valorization of the Marche Region (Marche Region, 2017a; hereafter "Development")
(7) Regional Forest Plan (Marche Region, 2005; hereafter "PFR")
(8) Regional Plan of forecasting, prevention and active fight against forest fires 2017–2019 (Marche Region, 2017b; hereafter "Forest Fires 2017–2019")
(9) Regional Forest Table (Marche Region, 2011; hereafter "Forest Table")
(10) Regulator Plan for Aqueducts (Marche Region, 2014; hereafter "Aqueducts").

These initiatives cover a wide range of sectors, addressing issues from waste management and rural development to forest planning and aqueduct regulation.

The strategies, issued between 2005 and 2019, prioritize in the majority of cases both adaptation and mitigation efforts (Table 7.3). They are embedded in regional laws and programs, covering the entirety of the Marche Region or specific areas within it, such as coastal zones (GIZC Plan) or forests (PFR, Forest fires 2017–2019, Forest Table).

Context

The general objective of the GIZC Plan is to facilitate the sustainable development of coastal areas through planning activities, in order to reconcile the economic, social and cultural development, in response to changes induced by the variation of the sea level, by extreme events, by anthropic impacts, while maintaining unchanged the functions of the coastal system for a longer period. This strategy foresees both maintenance interventions (existing rigid works, previous beach nourishment, restoration of sandy shores) and new interventions.

The overall objective of the PEAR 2020 strategy is to define the ways in which the Marche Region intends to comply with the mandatory provisions of the Ministerial Decree of March 15, 2012, regarding the adjustment of the percentage of renewable energy in gross final consumption and the revision of the Regional Energy Strategy until 2020, following the changes in the socio-economic context. A roadmap up to 2030 is presented with the aim of establishing a set of guidelines for the regional energy policy in terms of energy savings and efficiency, electricity generation, and the development and efficiency of energy network infrastructure.

The main objective of the Sustainability document is to contrast climate change and develop a new strategy for sustainable development. The strategy addresses sustainability development in line with the challenges posed by the global 2030 Agenda, referring to the economic, social, and environmental dimensions.

The overall purpose of the PRGR is to regulate waste management activities. The strategy is organized as follows: (1) Analysis and evaluation of the operational situation (for municipal and hazardous waste), identification of system criticalities and definition of planning objectives; (2) Definition of forecasting models, identification of strategic lines and proposed methodologies for achieving the planning objectives; (3) Drafting of the Regional Plan proposal.

Two are the main objectives of the Rural Strategy 2014–2022: (i) to promote sustainable rural development in the European Union (EU), and (ii) to develop the agricultural sector to achieve greater territorial and environmental balance, which also means greater competitiveness and innovation. The strategy is structured as follows (1) Innovation in agriculture and forestry; (2) Increasing the competitiveness of agriculture; (3) Organizing the agri-food supply chain and risk management; (4) Restoring ecosystems; (5) Efficient use of resources and a low-emission and climate-resilient economy; (6) Social inclusion and poverty reduction.

The Development strategy aims to recover the regional areas severely affected by a massive earthquake in 2016, through a series of linked and integrated destination marketing activities, synergizing with all regional tourism clusters and diversified on different markets and on different targets. The overall objective is to promote tourism. The Marche Region will directly promote its territory as a tourist destination by defining a new brand and investing, also with the support of a qualified international communication and marketing agency, in advertising campaigns to be carried out on television, radio and cinema, mass media in general and printed media.

The PFR strategy aims at the conservation of biodiversity and the protection and diffusion of agro-forestry systems of high public, socio-economic and landscape-environmental value. The main objectives are: protection and conservation of rare, exclusive and disappearing floristic species; preservation of the current vegetation structure in the mountains and high hills; preservation of the aesthetic and historical characteristics of those plant elements that characterize the regional environment; restoration, consolidation and development of the botanical and vegetation heritage for ecological and soil protection purposes.

The overall objective of the Forest Fires Strategy 2017–2019 is to support the activities of forecasting, prevention, and repression of violations related to forest fires. Forecasting and prevention activities include updating of water supply points and constant updating of significant territorial skills. The Prevention activities are planning and organization of reconnaissance actions, surveillance, sighting of doors by the State Forestry Corps Stations, on a day and night basis, in relation to the areas or greater risk and according to the available resources. The activities of repression of the infringements are organization of the investigative activity of the State Forestry Corps Stations, of the Investigative Units of the Forest and Environmental Police and the relative specialized personnel.

The main objective of the Forest Table is to implement an active sustainable management of forests and the forest sector, to ensure the natural renewal and protection of forest ecosystems and the socio-economic development of the same sector, to provide continuity and job security in the sector. The strategy is designed to systematically identify and promote rational, modern approaches involving interventions in forests. These actions will be supported by public resources to activate and implement sustainable, proactive management of forest resources by owners, entrepreneurs and managers, whether public, private or in partnership. Priority will be given to collaborative efforts aimed at joint management of large forest areas. The strategy includes the implementation of forest management practices aimed at reducing greenhouse gas emissions. It includes the development of comprehensive knowledge tools such as inventories and detailed forest plans to facilitate informed management of the multiple values and functions of forest resources. It also calls for the implementation of plans and policies in accordance with protocols, resolutions, conferences, guidelines and regulations governing the sector. The strategy emphasizes plans and actions aimed at preserving and enhancing landscapes that embody natural and cultural values, including human interventions and the conservation and enrichment of biodiversity. It also aims to promote a shared appreciation of the values of forest resources, including their role in shaping landscapes. It seeks to maximize the multifunctionality and public benefits of forests, including soil and water conservation, promotion of wood energy supply chains and other forest products, reforestation, tourism, public enjoyment and environmental education.

Finally, the main objective of the Aqueduct strategy is to identify the sources of supply to be "reserved" for drinking water, with projection to the year 2050, as a strategic resource of the region to be protected and preserved, to guarantee the community and future generations the availability of the primary public good. The strategy is organized around the following components: a comprehensive assessment of resources within a unified management framework for the territory; limiting total water losses to a maximum of 20%; optimizing the use of resources and implementing water conservation measures; integrating networks to ensure a balanced supply throughout the territory and in adjacent regions; exploring new water sources to replace poor quality ones and mitigate water deficits; and abandoning reliance on unreliable sources in favor of safe and reliable water sources that meet qualitative and quantitative standards.

The motivating factors behind the development of these strategies include several elements. Most of them are driven by the implementation of European policies and United Nations summits/reports, as well as by the impact of extreme weather events and the economic costs of inaction. The Sustainability, Development, and Forest Fires 2017–2019 strategies are also based on scientific research. Private sector interests motivate PSR 2014–2022, Development, and Forest Table, while examples from other countries/regions have been considered for Sustainability and PFR. Finally, Development also recognizes the possibility of new opportunities arising from the implementation of the strategy.

Relevance

The strategies outlined in the Marche Region integrate climate scenarios and projections, considering geographical specifics and the latest scientific knowledge.

For example, climate scenarios and projections have been studied both in the previous and in the current GIZC Plan. Moreover, the study of climate is fundamental in the PFR, because the typical climate of the Adriatic regions and climate changes influence the characteristic fauna of the Marche Region. In the Forest Table, the Marche Region has been divided into climatic areas, which identify areas with a specific type of climate and with specific plant species. In addition, for the Forest Fires 2017–2019, the climate change is evaluated, because the climatic conditions in the different seasons can influence the fire dynamics. In addition, the possibility of water supply must be taken into account in order to extinguish fires, as well as the prevailing wind direction. Finally, in the Aqueduct, the physical and climatic characteristics of the area where the drinking water is used, is defined by the scenarios of the IPCC Fifth Assessment Report Climate Change Assessment Report (2013).

These analyses are supported by a comprehensive assessment of vulnerabilities and climate risks. This is the case for the GIZC Plan, which considers that coastal erosion, extreme events and sea flooding are influenced by climate change. The management of catastrophic events is also considered in Sustainability, Development (earthquakes) and in Forest Fires 2017–2019 and Forest Tables (forest fires and geo-hydrological hazards). In particular, the development strategy is based on the 2016 earthquake, which was an exceptional event for the amount of damage it caused. Moreover, PEAR 2020 refers to the climate change risks identified by the 20–20–20 European strategy. Other elements of vulnerability considered are environmental (PRGR, PSR 2014–2022, PFR, Forest Fires 2017–2019, Forest Table, Aqueducts), economical (PSR 2014–2022, Aqueducts), social (PSR 2014–2022), and cultural (PRGR).

Almost all strategies explicitly consider transboundary impacts, with the exception of PEAR 2020. For example, the GIZC plan foresees the involvement of neighboring coastal regions (Abruzzo and Emilia-Romagna), while other strategies refer to the overarching legislation and its scope (e.g., national or European).

Several scientific organizations and institutions were explicitly involved in the development of the strategies, such as the National Agency for New Technologies, Energy and Sustainable Economic Development (ENEA) in the PEAR 2020. Other regional agencies, such as the Agency for Environmental Protection of the Marche Region (ARPAM) and the Veterinary Service, Hygiene, Safety and Nutritional Quality of Food gave a specific technical-scientific contribution in the PRGR and PSR 2014–2022, respectively. The Agency for Agro-food Sector Services of the Marche Region (ASSAM), the Regional Plant Health Service and the Order of Agronomists and Forest Doctors were also involved in the Regional Forest Table. Finally, the Polytechnic University of Marche participated in PSR 2014–2022, PFR, Forest Fires 2017–2019, Forest Table and Aqueducts.

The participation of organizations linking science and policy was confirmed in only a few cases, namely the PRGR, the PFR and the Aqueducts. Interestingly, the latter involved the regional section of the National Association of Italian Municipalities (ANCI).

Implementation

The 10 regional strategies cover the full range of sectors and vulnerabilities considered in the analysis, namely agriculture, biodiversity/ecosystem conservation, coastal management, communications, desertification and drought, electricity supply, emergency management, fisheries, floods and landslides, foreign policy, human health, industry, land use, liability and insurance, tourism and leisure, transport and infrastructure, urban settlements, and water resources management (Table 7.4). PRGR also includes the waste management. Interestingly, communication, electricity supply, emergency management, fishery, foreign policy, and industry are addressed by only one strategy, whereas biodiversity/ecosystem conservation, land use, responsibility and insurance, and urban settlements are addressed by most of the strategies.

In terms of timeframes, most of the strategies envisage "long-term" implementation, focusing on actions to be carried out over 5 years or until an update, with the exception of the GIZC Plan and Forest Fires 2017–2019, which envisage "medium-term" implementation.

In terms of measures, most strategies include "green" measures, in PRGR paired with "soft" measures. GIZC Plan includes both "gray" and "soft" measures, whereas Development and Forest Fires 2017–2019 include only "soft" measures.

Table 7.4 List of the vulnerable sectors and the Marche Region strategies that address them

Sector	Strategy engagement
Agriculture	PRGR, PSR 2014–2022, forest fires 2017–2019, Forest Table
Biodiversity/ecosystem conservation	GIZC plan, PRGR, PSR 2014–2022, PFR, forest fires 2017–2019, Forest Table, aqueducts
Coastal management	GIZC plan, PFR, forest fires 2017–2019
Communications	Development
Desertification and drought	PSR 2014–2022, forest fires 2017–2019
Electricity supply	PEAR 2020
Emergency and rescue services	Development
Fishery	PFR
Floods and landslides	GIZC plan, sustainability
Foreign policy	PEAR 2020
Human health	PEAR 2020, sustainability, PRGR, PSR 2014–2022, aqueducts
Industry	PEAR 2020
Land use	GIZC plan, PEAR 2020, sustainability, PRGR, PSR 2014–2022, PFR, forest fires 2017–2019, Forest Table
Responsibility and insurance	PEAR 2020, sustainability, PRGR, PSR 2014–2022, PFR, forest fires 2017–2019, Forest Table
Tourism and leisure	GIZC plan, development, forest fires 2017–2019
Transport and infrastructure	GIZC plan, sustainability
Urban settlement	GIZC plan, PEAR 2020, sustainability, development, PFR, forest fires 2017–2019
Water resource management	Sustainability, forest fires 2017–2019, aqueducts

Specifically, the GIZC Plan actions include maintenance of existing works such as previous nourishment, waste and stranded materials management; and structural interventions such as new nourishment, defense and sustainable development works.

The PEAR 2020 strategy includes the following: energy saving awareness and information campaigns and incentive programs characterized by bureaucratic simplicity; the use of renewable energies, with particular reference to wind energy and biomass of agro-forestry origin, also for the production of biofuels, and solar energy to be used in construction; energy eco-efficiency through strong and widespread action of technological and managerial innovation; and the production of distributed electricity and thermal energy in large areas of the Marche valleys and coasts.

The Sustainability strategy conveys measures regarding public transport (TPL), bicycle network, hydraulic risk mitigation and renewable energy and energy savings (reasoning with PEAR 2020).

The PRGR focuses on reducing the quantity, volume and hazardousness of waste and improving the technical/environmental performance of existing waste management facilities.

The PSR 2014–2022 includes a wide range of initiatives to support rural development. These include actions such as knowledge translation, farm management advisory services, quality improvement of agricultural and food products, investments in fixed assets, restoration of agricultural production after natural disasters, development of farms and businesses, renovation of rural villages, development of forest areas, creation of associations and producer organizations, agri-climate-environmental payments, promotion and maintenance of organic farming, compensation for environmental damage, animal welfare, forest protection, cooperative efforts, and support for participatory local development.

The Development strategy involves the cultural aggregation and networking of several different clusters within the Marche Region. These clusters include the supply and expertise of the tourism industry; the unique handicrafts and micro-enterprises specializing in characteristic products; the strengths and competitive advantages in key export sectors such as agri-food, footwear, leather goods, household appliances and electrical equipment; the capabilities of ICT companies, particularly those integrated with online projects; cultural and creative enterprises linked to these clusters; innovative start-ups and the sharing economy; and the expertise in high education and research of the region's four universities. This strategy aims to harness and connect the diverse strengths of these clusters, fostering collaboration and synergy between them to drive the region's economic growth and innovation across multiple sectors.

The PFR includes training and information programs for forest operators, acquisition of skills, use of extension services, and improvement of forest infrastructure to increase the economic value of forests and their products. The PFR also includes initiatives such as afforestation of agricultural land, implementation of agroforestry systems, compensation for Natura 2000, and environmental payments to support forests. In addition, it focuses on the restoration of forest potential, preventive measures and support for non-productive forest investments to ensure the conservation and responsible management of forest resources.

The Forest Fire prevention activity for 2017–2019 consists of the implementation of targeted actions to reduce the causes and potential ignition of fires, as well as interventions to mitigate the resulting damage. To this end, all the systems and means of control and monitoring of the risk areas will be used, as well as the cultivation measures suitable for improving the layout of the forest stands. The sustainable management of forest resources can play a key role in the transition to a low-carbon economy, in the carbon sequestration process and in the provision of environmental public goods. Economic and social needs are increasingly demanded by society, including the possibility of tourist-recreational use.

The Forest Table strategy is aimed at the following: forest cleaning and forest maintenance operations; silvicultural operations for fire-fighting purposes divided into individual categories; regional forests; regional forestry interventions with forest fire prevention purposes.

Finally, the Aqueducts is a project to regulate the pressure in the water network; to adjust the storage tanks; to connect the water network.

To implement these measures, most of the strategies envisage the use of either regulatory or economic/financial policy instruments, or both. PEAR 2020, Sustainability, and Development also use communication-related policy instruments and voluntary mechanisms for their implementation.

Interactions

The development and implementation of the strategies involve a wide range of public and private stakeholders from different sectors.

The GIZC Plan involved a concerted effort by academic institutions such as the Polytechnic University of Marche, University of Urbino and University of Camerino, along with collaborative contributions from the National Research Council (CNR). Their focus converged on the preparation of preliminary analyses and guidelines for the drafting of the plan.

The PEAR 2020 focused on the integration of various regional entities, including mountain communities to adopt the plan and ENEA to analyze the energy balance.

The Sustainability strategy focused on uniting the efforts of the Marche Region, municipalities, local authorities, business associations and civil society to seek economic resources, proper management of water resources, enhancement of river areas, and protection against hydraulic risks.

The PRGR involved the local population, waste management infrastructures, ecological stations, and the National Ecological Packaging (CONAI) to develop comprehensive waste management plans and solutions.

The PSR 2014–2022 strategy worked with farms, municipalities, rural villages, and agribusinesses to raise awareness of the program among stakeholders.

The Development strategy foresees a network of citizens, businesses, media, regional and local administrations as well as national and European authorities. For each of these "channels" of relationships and data flow, policies, content, marketing and communication tools will be developed, along with specific feedback monitoring methods.

The PFR has prioritized the protection of forest ecosystems among mountain communities, provinces, park authorities, municipalities, the Agency for Agricultural and Food Services of the Marche Region (ASSAM), the State Forestry Corps, and academic institutions such as the Polytechnic University of Marche. The Marche Region regulates the plan and finances the actions to be taken together with the Mountain Communities, the Provinces, the Park Authorities and the Municipalities. ASSAM gives the opinion of the Regional Phytosanitary Service. The State Forestry Corps monitors safety in the forest area. The Polytechnic University of Marche processes scientific study data.

The Forest Fires 2017–2019 strategy bridged emergency response and academic contributions, involving the Marche Regional Fire Department, the Regional Command of the State Forestry Corps, and the Polytechnic University of Marche to organize strategic planning. The Marche Region carried out prevention activities with the State Forestry Corps. The Polytechnic University of Marche assisted in the study of the territory, the environment and the climate.

The Forest Table involved a wide range of stakeholders, including provinces, mountain communities, academic institutions, forest services, cooperatives and forest companies. This strategy aimed to develop a comprehensive, multi-dimensional approach to sustainable forest management. Specifically, the Polytechnic University of Marche studies the territory, the climate and the characteristics of the forest; ASSAM and the Regional Plant Health Service control the plant species; the State Forestry Corps carries out monitoring actions; the administrations involved in the strategy apply it in their territories.

The Aqueduct Strategy recognized the importance of specialized expertise within the Marche Region. It recognized the key role of qualified personnel in the study of the territory, the availability of water and the development of the plan.

The communication approach is mainly "top-down", since participatory activities has been envisaged only in a few strategies.

Coherence

The strategies are deeply integrated into many existing sectoral policies and frameworks at different levels—international, European, national, and regional.

Some of them integrate all policy levels. For example, the GIZC Plan is integrated into Administrative Resolution of the Regional Council (DACR) n. 169 of 02/02/2005 and Regional Law 15/2004, which refers to the former Coastal Plan; Plan for the Geo-hydrological Management of the Regional River Basins (PAI); Regional Laws 31/2012, 48/2013 and DACR 100/2014; Directive 2007/60/CE; Legislative Decree 49/10; Legislative Decree 152/06; Law 221/2015. PEAR 2020 is integrated in the document of the "World Energy Group" drafted by the International Energy Agency; in the European strategy "Energy Union" and in the "National Action Plan for Renewable Energies of Italy" (PAN). The sustainability strategy resonates both with a Regional Law, the D.G.R. n. 225 of 09/02/2011, and with the European Agenda 2030. PRGR is included in the Italian Legislative Decree of April 3, 2006, n. 152, art. 199 and the Legislative Decree of December 3, 2010, n. 205, the Directive 2008/98/

EC and the Directive 2012/19/UE. The PSR 2014–2022 is included in the national strategic plans and in the European strategy 2020.

On the other hand, the Forest Fires 2017–2019 is integrated into the National Law of November 21, 2000, n. 353, art. 3.

The other strategies include regional policies: PFR and Forest Table refer to the Regional Law 23 February 2005, n. 6, art. 4; Aqueducts refers to the Decree of the Regional Council n. 238 of March 10, 2014.

Finally, development is not integrated in any other level of policy.

Efficiency

All strategies have specific timelines for implementation and are either underway or completed, with no deviations reported. The initial implementation dates range from 2005 to 2017, while the completion dates range from 2019 to 2050, but mostly 2020. The PFR and Forest Table are in effect until an updated law is enacted.

In all cases the funding is specified (EU, national, regional) and the allocated budget is reported. With a significant reliance on European Regional Development Funds (GIZC Plan), Italy's Development and Cohesion Fund (GIZC Plan and Sustainability), alongside contributions from private entities and municipalities, the Region secures varied funding ranging from €1.8 million for the maintenance of coastal defense works to €3.4 billion for the integrated management of coastal areas (both in the GIZC Plan). The Marche Region plays a central role, committing substantial amounts, including allocations from the European Regional Development Fund (ERDF), focusing in particular on specific sectors such as local transport, cycle paths and hydraulic risk mitigation (GIZC Plan and Sustainability).

Achieved Results

Evaluation of results is foreseen for most of the strategies, except for Sustainability. The evaluation methods consist mainly in monitoring, mostly by the Marche Region itself. The GIZC plan has set up extensive monitoring bodies in various fields: the Marche Region and the Polytechnic University of Marche are in charge of monitoring biodiversity, while national bodies such as the Statistical Institute (ISTAT) and other environmental agencies are responsible for monitoring water quality. Soil monitoring is the responsibility of the Institute for Environmental Protection and Research (ISPRA) in collaboration with the Marche Region, while cultural heritage is monitored jointly by the Marche Region and the Marche Superintendency of Archaeology, Fine Arts and Landscape. Climate change monitoring is the sole responsibility of the Marche Region, while ARPAM is entrusted with the crucial task of monitoring human health.

Other monitoring bodies are the State Forestry Corps in PFR and Forest Table and the Italian Renewable Sources Monitoring System (SIMERI) managed by the Energy Services Manager (GSE) and ENEA.

Key strengths identified in these strategies include the high-quality technology for robust analysis, ensuring a solid foundation for decision-making and implementation in the GIZC Plan and the contribution to the 2030 Agenda's objectives in the Sustainability strategy. PRGR emphasizes on maximizing material recovery and minimizing hazardous waste, coupled with the elimination of land-fill disposal for undifferentiated waste. The strength of PSR 2014–2022 strategy is the creation of a national rural network. The Development strategy capitalizes on the strong appeal of Made in Italy on international markets. The PFR and Forest Table emphasize the preservation of the richness of flora and therefore of fauna "Mediterranean forest" and also to the protection of the entire sector and the regional territory as a whole. The Forest Fires 2017–2019 strategy emphasizes sustainable actions on the territory.

On the other hand, the weaknesses identified include: the need for better integration of sustainable development principles, especially in the management of water resources (GIZC Plan); the lack of skilled employment opportunities and adequate energy infrastructure, which hinders the retention of younger generations in the region (PEAR 2020); resource constraints and timeliness issues affecting the implementation of the strategy (Sustainability); marginal waste recovery allowances (PRGR); the logistical difficulties in cooperation between agricultural systems (PSR 2014–2022) and economic struggles affecting profitability (Development); the management hurdles in disadvantaged areas, such as mountainous regions (PFR); logistical difficulties in covering a vast territory (Forest Fires 2017–2019); the delicate balance between protectionism and economic growth (Forest Table); and the challenge of securing sufficient funding for research initiatives (Aqueducts).

Specific lessons revolve around the significant role of cooperation as a cornerstone (PSR 2014–2022, Development, Forest Fires 2017–2019, Forest Table). It also emphasizes the need for careful planning calibrated to regional sensitivities and the criticality of real-time monitoring to avoid irreversible impacts (GIZC Plan). In addition, PEAR 2020 emphasizes the evolving renewable energy landscape, highlighting the growth of the photovoltaic market, the increasing importance of electrical biomass, and the irreplaceable nature of wind energy in the renewable energy sector. In particular, the PRGR recognizes the long evolution of waste management and the need for sustainable, adaptive processes. Finally, the PRGR emphasizes the essential coordination and planning required for active and sustainable vegetation management based on phytoclimatic influences.

7.2.2 Šibenik-Knin County

General Information

The Šibenik-Knin County has implemented one climate change adaptation strategy called the Integrated Coastal Zone Management Plan of the Šibenik-Knin County (Priority Actions Programme Regional Activity Centre, 2016; hereafter "ŠKZ"). The document refers to the Croatian guidelines for coastal planning of sub-national regions. The strategy was issued in 2016 by the Center for Regional Activities of the Priority Action Program in Split.

Context

The overall objective of the ŠKZ strategy is to provide guidelines for the integrated management of the coastal zone in the Šibenik-Knin County, taking into account climate change. Specifically it aims: to promote the acceptance of sustainability and resilience as criteria for the development of the coastal zone; to create conditions for the definition of sustainability in concrete areas; to contribute to the strengthening of participation and education; to provide guidelines for sectoral policies and plans for the achievement of sustainability and resilience; to make recommendations for the resolution of conflicting issues in the achievement of sustainability and resilience; to provide a platform for the sustainable development of the coastal zone of Šibenik-Knin County based on water as its basic resource, blue economy, and smart specialization. The document provides a description of the current state of coastal areas (population, coastal morphology, and water areas), identification of key problems and vulnerability of the coastal area. The important part of this document is the vision for future development and scenarios for achieving this goal. The Coastal Plan is not a "normative" plan, it belongs to a group of "indicative" plans that provide guidance for managing certain issues and defining solutions that will be incorporated into normative plans. The Plan is primarily addressed to the employees of the County Administration responsible for the County's coastal area, institutions and administrations whose activities are related to the coast. The Plan is also addressed to the general public, the population living in the coastal area or those who are there for professional activities, investments or leisure and recreation.

The motivating factors behind the development of these strategies include several elements. Most of them are driven by the implementation of European policies and United Nations summits/reports, as well as the impact of extreme weather events, scientific research and private sector interests.

Relevance

The ŠKZ strategy is based on climate scenarios and projections, using the United Nations Development Program (UNDP) Human Development Report for Croatia (2008): this report states that the County has coastal areas that are highly vulnerable to climate change. The strategy highlights the main vulnerabilities related to different sectors such as coastal protection, water services, agriculture and fisheries, nature conservation and tourism. Transboundary risks are explicitly considered in the socio-economic analysis: Adriatic-Ionian, central Europe, Danube, Mediterranean. Education and research institutions are considered as stakeholders through interviews, and civil society (NGOs) are involved as bridging organizations.

Implementation

In addition to coastal management, the ŠKZ strategy covers a wide range of issues, including agriculture, biodiversity/ecosystem conservation, fisheries, floods and landslides, land use, liability and insurance, tourism and recreation, and water resource management.

In terms of the timeframe, the program foresees a "long-term" implementation, i.e., more than 5 years.

The strategy includes "green" and "soft" measures. Specifically, it uses the Dynamic and Interactive Vulnerability Assessment (DIVA) method, which provides forecasts of climate impacts and costs that are used in the preparation of the Coastal Plan. The DIVA method builds modular, integrated computer models by bringing together knowledge from natural and social disciplines. DIVA allows users to select scenarios and adaptation strategies and to compare results for different scenarios and adaptation strategies. It is used to assess the potential costs of climate variability and change for the entire Croatian coast. Due to the small size of Šibenik-Knin County, DIVA results for the Croatian coastal zone will be downscaled for Šibenik-Knin County for the purposes of the Coastal Plan. This task will be carried out by a group of international experts, taking into account local data on tourism, agriculture, health, infrastructure and possible costs of forest fires. Climate impacts and possible adaptation measures were discussed with stakeholders in the county using a participatory method called Climagine, which was developed as part of a larger United Nations Environment Program—Global Environment Facility (UNEP/GEF) project.

Both regulatory and voluntary policy instruments will be used to implement these measures.

Interactions

Local, regional, and national institutions were actively involved in the development of the ŠKZ strategy: Ministries, municipalities, public institutions (national water and forest management companies, NGOs ...); government and policy-making bodies—thirteen interviews; regional and local authorities; state institutions represented in Šibenik-Knin County; educational and research institutions; civil society (NGOs); businesses. They were interviewed in order to acquire local knowledge, to elicit local creative ideas and to create a relationship of trust on which the future implementation of the Coastal Plan is based. The communication approach is mainly top-down.

Coherence

The ŠKZ plan is part of the United Nations Environment Program—Mediterranean Action Plan (UNEP/MAP) project "Integration of climatic variability and change into national strategies to implement the ICZM Protocol in the Mediterranean" (ClimVar and ICZM), and was also prepared following the planning steps suggested by the Guidelines for an Integrative Methodological Framework and the Guidelines for Adaptation to Climate Variability and Change along the Mediterranean Coast.

Efficiency

The timeframe of ŠKZ is between 2016 and 2030, so the strategy is in progress with no reported deviations.

The funding is not specified and the allocated budget will be calculated on the basis of the DIVA method.

Achieved Results

The evaluation of the results is planned through monitoring, modeling and forecasting. The evaluation will be carried out by the Department of Environmental Protection and Public Utilities and the Office for Spatial Planning (Šibenik-Knin County).

The main strengths of the ŠKZ were identified as: stakeholder participation; establishing a link between relevant sectoral plans, spatial plans, the Development Strategy of the Šibenik-Knin County and the Coastal Plan; collection and storage of relevant information, knowledge and awareness on climate variability and change and on the possible impacts on the coastal zone. On the other hand, the budget and precise timetable for implementation are not provided. Ambiguity regarding the legal basis for the preparation of the Coastal Plan, since the relevant planning legislation in Croatia does not provide for the preparation of a Coastal Plan at the county level; the existing spatial and physical plans do not give special attention to the coastal zone; the issue of climate change is still very new for the stakeholders and inhabitants, so they may not fully grasp the importance of the issue, resulting in delayed measures to adapt to climate change and minimize its negative impacts; difficulties in obtaining relevant information.

Finally, the main lessons learned are: be able to identify key issues; apply a holistic approach; territorial coverage must also be defined; integration is easy to recommend but difficult to achieve; a coastal plan offers a series of recommendations but also opens a series of questions.

7.3 Southern Adriatic Strategies

This section examines the current mitigation and adaptation strategies implemented in the Sothern Adriatic area, specifically focusing on the Apulia Region for the Italian side and on the Dubrovnik-Neretva County for the Croatian side.

7.3.1 Apulia Region

General Information

The Apulia Region has implemented a diverse array of strategies addressing climate change adaptation and mitigation across various sectors (Table 7.5):

(1) Apulia Regional Operational Program (Apulia Region, 2015a; hereafter "POR Apulia 2014–2020")
(2) Operational Fire Danger Prevention Platform (Interreg Greece-Italy OFIDIA, n.d.; hereafter "OFIDIA")

Table 7.5 List of strategies issued in the Apulia Region, including the indication of the type of strategy and the year of issue

Strategy	Mitigation	Adaptation	Year
POR Apulia 2014–2020	X	X	2015
OFIDIA		X	2015
Rural development		X	2015
Waste	X		2017
SME	X		2008
Housing	X		2008
Pollution	X		2006

(3) Regional Rural Development Program (Apulia Region, 2015b; hereafter "PSR 2014–2022")
(4) Recovery and reuse of surplus, food waste and pharmaceutical products (Apulia Region, 2017; hereafter "Waste")
(5) Aid for investments of Small and Medium Enterprises in energy saving, in high-efficiency cogeneration and in the use of renewable energy sources (Apulia Region, 2008a; hereafter "SME")
(6) Standards for sustainable living (Apulia Region, 2008b; hereafter "Housing")
(7) Urgent measures to reduce light pollution and save energy (Apulia Region, 2006; hereafter "Pollution").

These initiatives cover a wide range of sectors, addressing issues from waste management and rural development to sustainability and energy savings.

The strategies, adopted between 2006 and 2017, prioritize mitigation in most cases, adaptation in fewer, and in one case a dual objective combining both (Table 7.5). They are mostly embedded in European and regional directives or regulations and in cooperation programs between Member States (Italy-Greece in OFIDIA). They cover the whole of the Apulia Region or specific municipalities within it, such as Bari, Brindisi and Lecce in OFIDIA.

Context

In the POR Apulia 2014–2020, the Region has chosen to achieve a strong integration between the European Regional Development Fund (ERDF) and the European Social Fund (ESF) by defining a multi-fund program. The POR Apulia 2014–2020 aimed to continue the previous actions implemented under the 2007–2013 programming, in line with the priorities and objectives of the Europe 2020 strategy. It focused on four key areas: (1) Research and Innovation, aiming at industrial research, strengthening of the digital system, competitiveness and internationalization; (2) Contextual Policies, aiming at energy efficiency, territorial security, cultural and environmental preservation, sustainable transport and revitalization of peripheral areas; (3) Labor Market, Social Inclusion and Welfare, prioritizing employment increase, poverty reduction, social inclusion and educational improvements; (4) Strengthening Administrative Capacity, emphasizing skills enhancement, simplification of bureaucracy, promotion of transparency and use of common intervention methods.

The overall objective of the OFIDIA strategy is to protect the environment and conserve natural resources through an environmental monitoring platform and fire risk forecasting for the Apulia Region and northwest Greece. The project involved the installation of automatic visual and thermal cameras throughout the forested area and the development of an ICT infrastructure that provides online geographic maps with advanced weather forecasts, fire risk indices and historical information on meteorological sensors, forecast data and forest fires.

The main objectives of the PSR 2014–2022 are: (1) Facilitate knowledge transfer and innovation in agriculture, forestry and rural areas, while improving farm profitability, agricultural competitiveness and sustainable forest management through innovative technologies. (2) Promote the organization of the food chain, including the processing and marketing of agricultural products, animal welfare and risk management within the sector. (3) Preserve, restore, and enhance ecosystems related to agriculture and forestry. (4) Promoting resource efficiency and the transition to a low-carbon, climate-resilient economy in agriculture, food and forestry. (5) Focusing efforts on social inclusion, poverty reduction and economic development in rural areas. A justification of the choice of needs to be addressed by the PSR and of the choice of objectives, priorities, specific aspects and target setting is based on the Strengths, Weaknesses, Opportunities, and Threats (SWOT) analysis and the needs assessment. Where appropriate, a justification of the thematic sub-programs will be included in the program. The justification shall demonstrate compliance with the requirements of Article 8, paragraph 1, letter (c), points (i) and (iv) of EU Regulation n. 1305/2013.

The overall goal of the waste strategy is to reduce the environmental impact of food waste and expired medicines by recovering and redistributing surplus goods, minimizing the amount of waste and promoting a circular economy that reuses unused items in a framework of solidarity. This includes raising awareness among food and pharmaceutical stakeholders and consumers, coordinating distribution efforts, fostering collaborative partnerships among industry players and donors, and overseeing waste reduction phases in partnership with healthcare companies to ensure product hygiene and safety standards. In addition, the promotion of transparent use and traceability of surplus food and medicines is in line with the directives of Law 166/2016, ensuring compliance. The Apulia Region, in order to protect the most vulnerable population groups and to encourage the reduction of waste, recognizes, encourages and promotes the activity of solidarity and charity aimed at the recovery and redistribution of surplus food and pharmaceutical products in favor of people in a state of poverty or severe social unease. The Apulia Region also promotes actions aimed at the recovery and redistribution of food waste for human and animal consumption.

The SME strategy aims at granting environmental aid to small and medium-sized enterprises for investments aimed at improving the environmental-energy balance of the company. The Apulia Region prioritizes sustainable development as a key driver to promote employment growth. This approach focuses on improving energy efficiency, increasing the use of renewable energy sources and promoting energy conservation. Despite the crucial role that small and medium-sized enterprises (SMEs) play in creating jobs, ensuring social stability, and driving economic vitality, they face challenges in implementing investments aimed at protecting the environment and improving quality. These hurdles often stem from potential increases in overall production costs, which can act as a disincentive for SMEs to pursue environmental improvement initiatives.

The Housing strategy aims to promote environmental sustainability and energy savings both in territorial and urban transformations and in the construction of public and private buildings. It defines the tools, techniques and construction methods that are sustainable in the instruments of territorial management, in new construction, building and urban recovery, and urban redevelopment.

Finally, the main objective of the Pollution Strategy is the reduction of light pollution and the resulting energy consumption, in order to preserve and protect the natural environment, both inside and outside protected natural areas.

This regulation implements the Regional Law of November 23, 2005, n. 15, with which the Apulia Region pursues the objectives of protecting environmental values aimed at the sustainable development of the regional community.

The motivating factors behind the development of these strategies include various elements. Most of them are driven by the implementation of European policies and private sector interests. Secondly, they are drawn from scientific research, impact of extreme weather events and recognition of new opportunities. Finally, POR Apulia 2014–2020 is also based on United Nations summits/reports.

Relevance

Apulia strategies rely heavily on climate scenarios and projections, such as those provided by the Natura 2000 network (POR Apulia 2014–2020), with the exception of the waste and pollution strategies. These analyses generally take into account geographical specificities and are supported by a comprehensive assessment of vulnerabilities and climate risks. In particular, the POR Apulia 2014–2020 "Measures to reduce hydrogeological risks and coastal erosion" (Axis V—Action 5.1.) foresees activities for the prevention and management of previously analyzed geo-hydrological and coastal erosion risks.

However, none of the strategies explicitly consider transboundary risks and thus allow for consultation with neighboring regions or states.

Scientific organizations and institutions were invited to provide input or comments on the development of only two strategies. POR Apulia 2014–2020 involved the National Research Council, the Universities of Bari, Salento and Foggia, and the National Institute of Geophysics and Volcanology.

Organizations linking science and policy have been reported POR Apulia 2014–2020 and the PSR 2014–2022. In particular, the first involved many entities such as the Agency for Environmental Protection of Apulia Region, the National Association of Italian Municipalities (ANCI), professional orders, representatives from the agricultural sector (e.g., CIA Apulia, Coldiretti Apulia, Confederation of Agricultural Producers), trade unions (e.g., CGIL, CISL, UIL), environmental and volunteer NGOs (e.g., WWF, F.A.I.). The second involved local action groups (GAL).

Implementation

The 7 regional strategies cover a wide range of sectors and vulnerabilities, namely agriculture, biodiversity/ecosystem conservation, coastal management, communications, desertification and drought, electricity, emergency and rescue services, fisheries, floods and landslides, foreign policy, human health, industry, land use, liability and insurance, tourism and leisure, transport and infrastructure, urban settlements, and water resources management (Table 7.6). Interestingly, all strategies address human health, while coastal management, communication, fisheries, foreign policy, industry, liability and insurance, and tourism and leisure are addressed by only one strategy (within the POR Apulia 2014–2020, OFIDIA, PSR 2014–2022 and SME).

In terms of timeframe, most of the strategies envisage "long-term" implementation, focusing on actions to be carried out over 5 years, with the exception of the OFIDIA, which envisages "medium-term" implementation.

In terms of measures, most strategies include only "gray" measures, except for POR Apulia 2014–2020, which also includes "green" and "soft" measures. On the

Table 7.6 List of the vulnerable sectors and the Apulia Region strategies that address them

Sector	Strategy engagement
Agriculture	POR Apulia 2014–2020, OFIDIA, PSR 2014–2022, SME
Biodiversity/ecosystem conservation	POR Apulia 2014–2020, PSR 2014–2022, pollution
Coastal management	POR Apulia 2014–2020
Communications	OFIDIA
Desertification and drought	OFIDIA, PSR 2014–2022
Electricity supply	POR Apulia 2014–2020, housing, pollution
Emergency and rescue services	POR Apulia 2014–2020, OFIDIA
Fishery	SME
Floods and landslides	POR Apulia 2014–2020, PSR 2014–2022
Foreign policy	POR Apulia 2014–2020
Human health	POR Apulia 2014–2020, OFIDIA, PSR 2014–2022, waste, SME, housing, pollution
Industry	POR Apulia 2014–2020
Land use	POR Apulia 2014–2020, PSR 2014–2022
Responsibility and insurance	PSR 2014–2022
Tourism and leisure	POR Apulia 2014–2020
Transport and infrastructure	POR Apulia 2014–2020, PSR 2014–2022, SME, pollution
Urban settlement	POR Apulia 2014–2020, housing, pollution
Water resource management	POR Apulia 2014–2020, PSR 2014–2022, SME, housing

other hand, the PSR 2014–2022 includes only "green" measures, while the waste strategies include only "soft" measures.

Specifically, the POR Apulia 2014–2020 includes various actions outlined under the 13 priority axes envisaged: Research, Innovation and Technological Development (Priority I): Strengthening the research infrastructure, promoting business investment in innovation and improving the regional innovation system. ICT accessibility and quality improvement (Priority II): Expanding broadband coverage, promoting the digital economy, and supporting ICT applications for various sectors. SME competitiveness (Priority III): Facilitating entrepreneurship, assisting SMEs in internationalization and innovation processes, and supporting product/service development. Sustainable energy and quality of life (Priority IV): Promoting energy efficiency, the use of renewable energy and low-carbon strategies in the public and housing sectors. Climate change adaptation and risk management (Priority V): Investing in disaster management systems and building resilience. Environmental protection and natural/cultural resources (Priority VI): Investing in the waste and water sectors, preserving natural/cultural heritage, promoting biodiversity and improving the urban environment. Transport systems and network infrastructures (Priority VII): Improving regional mobility, developing sustainable transport systems and improving rail and maritime transport. Employment sustainability and occupational mobility (Priority VIII): Promoting employment of specific groups, promoting job placement and improving the efficiency and quality of employment services. Social inclusion and the fight against poverty (Priority IX): Combating poverty, increasing employability, strengthening the social economy and improving health and social services. Investing in education, training and lifelong learning (Priority X): Preventing early school leaving, ensuring access to quality education, improving higher education and adapting training to the needs of the labor market. Strengthening institutional capacity (Priority Axis XI): Investing in public administration efficiency and good governance. Sustainable urban development (Priority XII): Focusing on urban renewal and ecological upgrading of production areas to improve living conditions and reduce environmental pressures. Technical assistance (Priority XIII): Improving the efficiency, effectiveness and quality of interventions, as well as verification and control mechanisms.

In the OFIDIA strategy, the early detection system uses automated visual and thermal cameras throughout the forest area. This equipment transmits information by radio to one of the five latest-generation central control rooms that the project has set up in Lecce, Brindisi, Bari and Ioannina (Greece), where it is used to coordinate the various closure operations.

The PSR 2014–2022 is multifaceted: it includes initiatives for vocational training and skills acquisition, supports the development of agricultural and forestry infrastructure, promotes the establishment of young farmers, improves broadband infrastructure, and supports reforestation and the resilience of forest ecosystems. It also promotes innovation in the agri-food and forestry sectors, encourages collaboration among small operators for shared work processes and tourism services, and facilitates joint actions for climate change mitigation and sustainable biomass production. The program also emphasizes non-productive investments aligned with environmental goals, animal welfare, and the development of new practices and technologies.

The Waste strategy includes a range of measures aimed at the recovery, redistribution and responsible management of food and pharmaceutical surpluses and food waste. It includes strategies to redistribute surplus food to the poor, reuse inappropriate food for animal consumption or composting, and develop educational programs to promote responsible consumption habits and reduce waste. The plan includes purchasing necessary goods and services for efficient recovery and storage, establishing storage and distribution centers equipped with facilities such as cold rooms, and creating local networks between relevant agri-food and third sector entities. It also promotes IT systems and collaboration to facilitate surplus recovery, encourages collaboration between food sector entities for responsible sales, supports citizen-led food sharing initiatives, and advocates for environmental protection against the negative impacts of food waste.

The SME strategy supports energy conservation initiatives, renewable energy production and high-efficiency cogeneration projects. Support is determined based on the eligible costs of the project, capped at certain percentages for diverse types of investments. For energy saving measures, the aid is limited to 25% for medium-sized enterprises and 35% for small enterprises. Meanwhile, investments in renewable energy sources and cogeneration can be supported up to 50% for medium-sized enterprises and 60% for small enterprises.

The proposed measures in the Housing strategy include various aspects of environmental sustainability within the tools of territorial governance, water conservation, energy supply efficiency, criteria for building materials, and certification of building sustainability. Territorial governance aims to achieve harmonious development, preserve cultural identity and mitigate environmental impacts by rehabilitating degraded areas and optimizing existing structures. Water conservation includes monitoring network efficiency, setting consumption standards, promoting natural purification, and using gray and rainwater systems. Energy efficiency focuses on defining consumption standards, using renewable energy sources, implementing cogeneration systems, and promoting environmentally friendly building materials. Building sustainability certification involves an independent assessment to certify a building's environmental standards, taking into account factors such as local sourcing, recyclability, low energy consumption, and health impact on occupants.

Finally, Pollution envisages the reduction of light pollution and nuisance lighting, as well as energy savings throughout the regional territory, through the rationalization of public and private outdoor lighting systems; the improvement of construction features and efficiency of lighting systems; the unification of design criteria to improve lighting quality and road safety through careful lighting design to guarantee energy and cost savings; the evaluation of lighting systems for green areas in urban areas; the prohibition of the installation of high-power public lighting systems that can disturb fauna in all areas of natural vegetation.

To implement these measures, most strategies envisage the use of both economic/financial and communication policy instruments. OFIDIA and SME only use economic/financial policy instruments for their implementation. The exception is the POR Apulia 2014–2020, which foresees all types of mechanisms: regulatory, economic/financial, voluntary and communication.

Interactions

A wide range of public and private stakeholders from different sectors are involved in the development and implementation of the strategies.

In the POR Apulia 2014–2020, Municipalities, private companies and citizens work together to promote smart, sustainable and inclusive growth in the region. They focus on innovation, resource efficiency, employment and social well-being. Public and private technicians work together to promote smart, sustainable and inclusive growth. Public institutions engage citizens in social development.

In OFIDIA, Municipalities (Bari, Brindisi, Lecce), Civil Protection (Italian and Greek) all aim for effective scientific collaboration between research institutions for the benefit of the local population.

PSR 2014–2022 beneficiaries, professional organizations, local authorities are responsible for promoting social inclusion, gender equality and non-discrimination. They engage in projects, environmental protection and promotion of social inclusion. Technicians, both public and private, engage citizens through public institutions to promote social welfare.

In the Waste strategy, the Apulia Region district, associated municipalities, private bodies focus on social and territorial development, agricultural practices and societal well-being through education, employment and resource management. The emphasis is on civic and solidarity roles in interventions against food waste.

In the SME strategy, small and medium enterprises and citizens work together to address technical and professional responsibilities for social progress.

The Housing strategy involves municipalities, citizens, construction companies and research institutions to engage in scientific cooperation for regional benefit and sustainable development.

In the Pollution strategy, Metropolitan City, Regional Parks, citizens, astronomical observatories and associations work together to identify and reclaim areas affected by light pollution, to mitigate the environmental impact of light pollution, ensuring careful evaluation of lighting systems for minimal disruption to biodiversity.

The communication approach is top-down only.

Coherence

The strategies are well integrated into numerous existing sectoral policies and frameworks at different levels—mostly European and national.

For example, the POR Apulia 2014–2020 aimed at ensuring continuity with the actions implemented within the POR programming 2007–2013, with the EU cohesion policy and with the European 2020 objectives. The waste strategy is integrated in the waste management policy, in key of the realization of a circular economy for a social solidarity, while the SME strategy is embedded in the growth and expansion policy of the employment base. On the other hand, the Housing strategy is integrated in new existing sectoral policies; they are Directive 2002/91/EC of the European Parliament and of the Council of December 16, 2002, on the energy performance of buildings and in line with Directive 2006/32/EC of the European Parliament and of the Council. Finally, the environmental strategy is integrated in numerous laws for the implementation of the National Energy Plan. For example, National Law n. 9 of January 9, 1991 (Rules for the implementation of the new National Energy Plan: institutional aspects, hydroelectric and power lines, hydrocarbons and geothermal energy, self-production and fiscal provisions) and National Law of January 9, 1991, n. 10 (Rules for the implementation of the National Energy Plan on the rational use of energy, energy saving and development of renewable energy sources).

Efficiency

All strategies have specific timelines for implementation and are either underway or completed, with no reported deviations.

The initial implementation dates range from 2006 to 2017, while the completion dates range from 2015 to 2023. Exception is the SME strategy for which the foreseen initial-end dates of implementation of each measure depend on terms set by the specific calls.

In some cases, the funding is specified (EU, national, regional) and the total budget is reported. For example, in the POR Apulia 2014–2020 the budget is €4.5 billion (between FESR and ERF) and in the PSR 2014–2022 the budget is €1.6 billion (European Agricultural Fund for Rural Development—EAFRD). In addition, €1.3 million is the budget allocated to OFIDIA from the Operational Program "Regional Research and Innovation" for the programming period 2007–2013. Finally, €600,000 is the budget for the fiscal year 2017 of the waste strategy from regional special financing.

Achieved Results

Evaluation of results has not been explained for most of the strategies, except for the POR Apulia 2014–2020 and Waste. However, the bodies responsible for monitoring and reviewing the strategies are indicated and seem to be mainly under the responsibility of the Apulia Region. The Region itself, together with specific departments such as the Regional Office for the Environment and Environmental Planning, leads these efforts in OFIDIA, Waste, SME, and Pollution. In addition, the Pollution Strategy works with expert committees, professional associations, and universities to ensure consistent application of environmental regulations. Municipalities also play a role in monitoring and verifying the implementation of interventions in the Housing strategy, working in agreement with the Region to ensure compliance with project results. Furthermore, in the POR Apulia 2014–2020, there are authorities overseeing financial management, control and audit aspects (MIRWEB and MIR system), with external monitoring involved in the final beneficiary monitoring.

Key strengths identified in the POR Apulia 2014–2020 include incentives for small and medium-sized enterprises (SMEs), promoting economic growth while taking advantage of a rich spectrum of environmental, cultural and human resources. In addition, POR Apulia emphasizes the use of high-tech instruments, which are expected to ensure efficient monitoring and implementation. Legislative measures to address environmental concerns, such as laws to curb emissions and initiatives to improve air quality in specific districts, stand out in the PSR 2014–2022. In addition, the SME strategy facilitates investments in environmental protection by offering increased aid intensity upon completion within specified timeframes. Finally, the Housing strategy promotes construction and renovation projects by reducing municipal taxes, secondary urbanization fees and construction costs, and encourages compliance with planning instruments.

On the other hand, the weaknesses identified in the POR Apulia 2014–2020 include several critical areas: problems related to the settlement, social and environmental degradation in urban suburbs; severe housing emergencies, especially in cities; insufficient green spaces and services; depopulation and demographic decline in inland areas. In addition, OFIDIA's weaknesses include shortcomings in accurately forecasting weather conditions. In the PSR 2014–2022, the lack of historical air quality data for certain pollutants hinders the assessment of the plan's effectiveness. The failure to reduce emissions from fossil fuel-based energy production, despite an increase in renewable energy production, and an increase in pollutants from industrial activities further undermine the effectiveness of the strategy. There are also concerns in the SME sector about the balance between tax incentives and the time required for a return on investment, highlighting potential financial constraints.

Interestingly, specific lessons learned or emerging needs/opportunities are explicitly outlined. For example, in OFIDIA, the lesson learned is that by protecting the forests of the regions of Apulia and Epirus from fires, the project contributes to a sustainable quality of life and economic prosperity in the cross-border region. In addition, in the PSR 2014–2022 the studied reports derived from the PSR 2007–2013 have been used to plan the strategy for the period 2014–2020. The housing strategy consists in the conscious use of energy resources and the development of sustainable building.

7.3.2 Dubrovnik-Neretva County

General Information

The Dubrovnik-Neretva County has implemented one climate change mitigation strategy named Environmental protection program for Dubrovnik-Neretva County for the period 2018–2021 (Dubrovnik-Neretva County, 2018; hereafter "DNZ"). The strategy has been issued in the year 2018 by the Dubrovnik-Neretva County itself.

Context

The general objective of the DNZ program is to provide guidelines for coastal planning and management in order to preserve the environment and enable sustainable development. The program analyzes the current status in different fields of interest (population, economy ...) and defines the main stakeholders in environmental protection. The focus is on many sectoral units (energy, agriculture, fisheries ...) with specific tasks to provide measures in the field of air protection, ozone layer and climate change mitigation, water protection, soil protection, sustainable management of the marine environment and coastal areas, preservation of biodiversity, and environment.

The motivating factors behind the development of this strategy include the implementation of European and national policies.

Relevance

The DNZ strategy does not use climate scenarios and projections to inform its approach. It does not conduct comprehensive vulnerability assessments and does not explicitly consider transboundary risks that cross administrative boundaries. In addition, scientific or bridging organizations/institutions are not explicitly involved in the development of the strategy.

Implementation

The DNZ strategy covers a wide range of issues, namely agriculture, biodiversity/ecosystem conservation, coastal management, electricity, fisheries, human health, industry, land use, tourism and leisure, transport and infrastructure, and water resources management.

In terms of timeframe, the program foresees a "medium-term" implementation, focusing on actions to be carried out in less than 5 years, i.e., from 2018 to 2021.

The strategy includes only "soft" measures. Specifically, two main measures have been identified: (i) implement fire protection measures in accordance with the DNZ Fire Protection Plan and the annual implementation plans for improving the DNZ fire protection; (ii) train, educate and regularly conduct exercises or readiness checks in accordance with the plans. Informing and involving the public as a whole in the plans and preparing for the implementation of the plans.

The use of either regulatory or economic/financial policy instruments is envisaged to implement these measures.

Interactions

In developing and implementing the strategy, the DNZ envisages the involvement of representatives of cities and municipalities, county departments, and other services to define stakeholders and actions for each sector. The communication approach is mainly "top-down".

Coherence

The strategy has been integrated into the Energy Development Strategy of the Republic of Croatia until 2030 with a view to 2050 (2021), which is the national climate change mitigation strategy aimed at realizing the low-carbon energy vision and ensuring the transition to a new period of energy policy, which will ensure affordable, safe and high-quality energy supply without additional burden on the state budget within the framework of state subsidies and incentives.

Efficiency

The timeframe of the PGZ was between 2018 and 2021, so the strategy is completed with no reported deviations. Currently, the Dubrovnik-Neretva County Environmental Protection Program for 2022–2025 is being prepared.

The funding is a co-financing from the European Union and the National Fund for Environmental and Air Protection, and the allocated budget is more than €5 million per year.

Achieved Results

The evaluation of the results is foreseen through monitoring actions. The evaluation will be carried out by the County Administrative Department for Communal Affairs and Environmental Protection.

The main strengths of the DNZ were identified as the assessment of the overall current situation in Dubrovnik-Neretva County with defined measures through different sectors. On the other hand, the main weakness of the program is that it does not consider future climate scenarios. In order to tackle the upcoming climate changes, any entity should issue a program that is interdisciplinary with relevant future scenarios and measurable impact of defined measures.

7.4 Summary

The regional strategies analyzed reflect the respective European and national regulations for mitigation and adaptation to climate change (Fig. 7.1). In the north Adriatic, the Friuli-Venezia Giulia Region focused on different topics, such as burden sharing, energy plans and environmental impact assessments. These 13 initiatives, introduced between 2007 and 2018, emphasized either mitigation (5) or adaptation (5) and were implemented through regional plans and government decrees. The County of Primorje-Gorski Kotar implemented one climate change mitigation strategy in 2019. In the central Adriatic, Marche Region adopted 10 strategies between 2004 and 2017, most of which focused on both mitigation and adaptation (6), targeting different sectors such as coastal zones and forests. Šibenik-Knin County followed the Croatian guidelines for coastal planning and adopted one strategy in 2016. In the south Adriatic, Apulia Region implemented 7 strategies between 2006 and 2017, focusing mainly on mitigation efforts (4) in sectors such as rural development and energy saving initiatives. Dubrovnik-Neretva County introduced one climate change mitigation strategy in 2018, with a three-year focus on environmental protection.

Across these regions, strategies vary in their reliance on climate scenarios and projections (Fig. 7.2), and in their assessing of vulnerabilities and risks (Fig. 7.3), and specifically cross-border risks (Fig. 7.4). In Friuli-Venezia Giulia Region, comprehensive analyses utilizing various strategies like Climate Impacts and PAI incorporate detailed forecasts and models, partnering with organizations and considering transboundary impacts. Similarly, in Primorje-Gorski Kotar County, although not directly

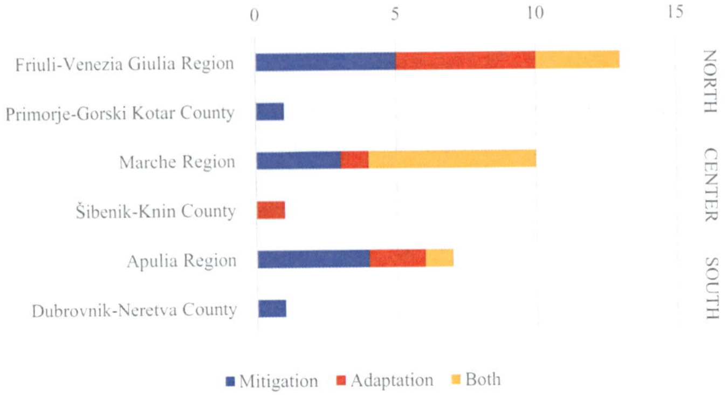

Fig. 7.1 Number and type of strategies by region and by Adriatic area (north, center, south)

involving scientific institutions, the PGZ strategy integrates geographical specifics and vulnerability assessments. In Marche Region, strategies extensively integrate climate scenarios, engage scientific institutions, and address transboundary impacts across various plans. Šibenik-Knin County's strategy, referencing UNDP reports, assesses transboundary risks, and involves education institutions and civil society. Apulia Region relies heavily on climate projections for most strategies but lacks explicit consideration of transboundary risks, engaging scientific institutions in a limited capacity. However, in Dubrovnik-Neretva County, the DNZ strategy notably lacks the use of climate scenarios, comprehensive vulnerability assessments, or direct involvement of scientific or bridging organizations.

Across the regions, the strategies aim to address multifaceted challenges (Fig. 7.5). In Friuli-Venezia Giulia, the main sectors covered are agriculture, biodiversity/ecosystem conservation, electricity supply, floods and landslides, and transport and infrastructure, with a focus on "long-term" and "gray" measures. However, there

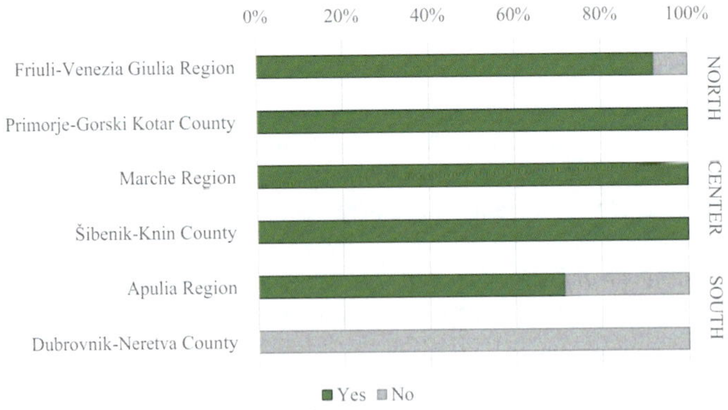

Fig. 7.2 Percentage of strategies relying on climate scenarios and projections by Adriatic area (north, center, south)

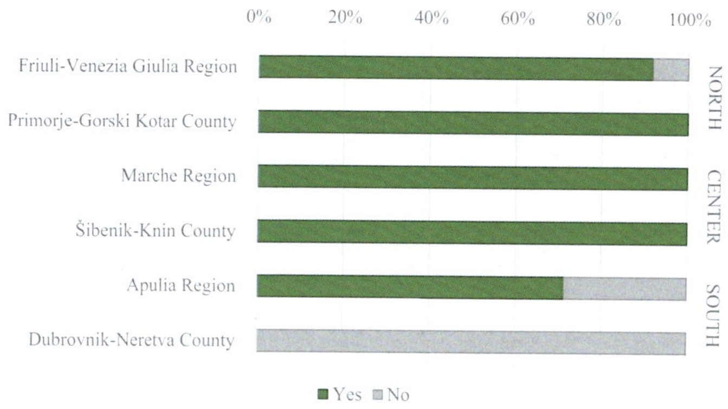

Fig. 7.3 Percentage of strategies assessing vulnerabilities and climate risks by Adriatic area (north, center, south)

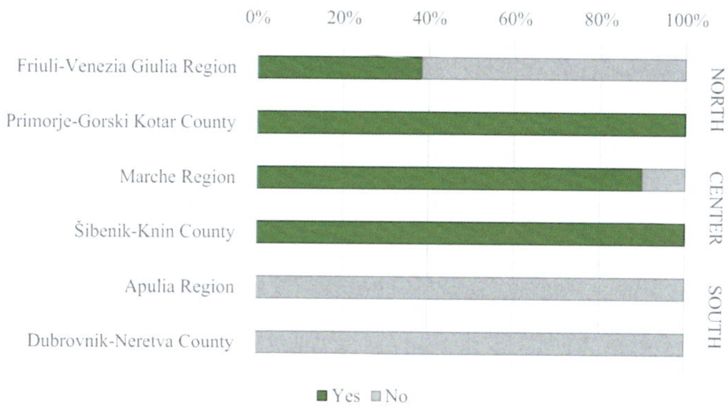

Fig. 7.4 Percentage of strategies assessing cross-border risks by region and by Adriatic area (north, center, south)

are gaps, with some sectors, such as emergency services and foreign policy, lacking specific strategies. In the County of Primorje-Gorski Kotar, the strategy focuses primarily on industry and transport, with a "medium-term" implementation plan using "soft" measures. Similarly, the strategies of the Marche Region cover various sectors, mainly biodiversity/ecosystem conservation, land use, liability and insurance, and urban settlement, with mostly "long-term" actions, and most strategies include "green" measures. The comprehensive strategy of Šibenik-Knin County covers agriculture, biodiversity/ecosystem conservation, fisheries, floods and landslides, land use, liability and insurance, tourism and recreation, and water management, and aims at "long-term" implementation with "green" and "soft" measures. The strategies of the Apulia Region cover a wide range of sectors, with human health in

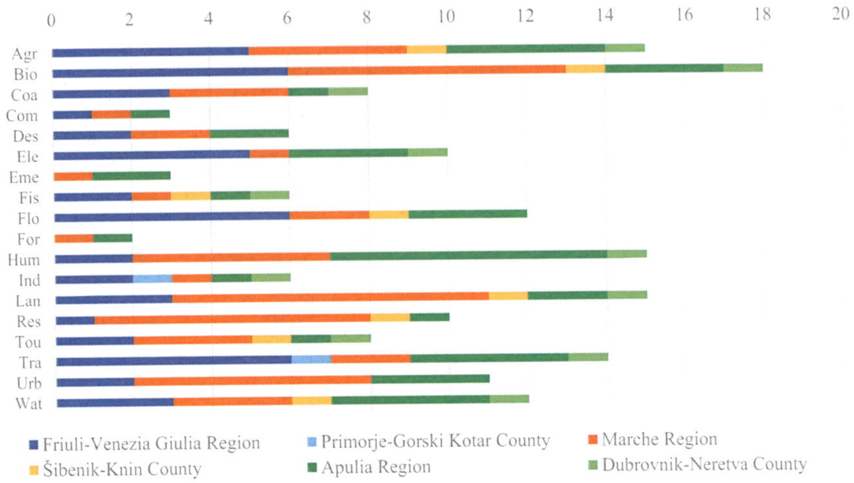

Fig. 7.5 Number of regional strategies by vulnerable sector considered (Agr: Agriculture; Bio: Biodiversity/ecosystem conservation; Coa: Coastal management; Com: Communications; Des: Desertification and drought; Ele: Electricity supply; Eme: Emergency and rescue services; Fis: Fishery; Flo: Floods and landslides; For: Foreign policy; Hum: Human health; Ind: Industry; Lan: Land use; Res: Responsibility and insurance; Tou: Tourism and leisure; Tra: Transport and infrastructure; Urb: Urban settlement; Wat: Water resource management)

all of them and agriculture, transport and infrastructure, and water resources management in most others, emphasizing "long-term" actions, with various measures, especially "gray", depending on the strategy. Finally, the Dubrovnik-Neretva County Strategy covers a wide range of sectors with a "medium-term" focus and exclusively "soft" measures.

Despite the differences in sectoral coverage, common features emerge, such as a focus on "long-term" planning, except for Primorje-Gorski Kotar County and Dubrovnik-Neretva County, and mostly "gray" or "green" types in Italy, while the "soft" type is observed in Croatia.

The strategies of these regions use a mix of regulatory, economic, communication and voluntary policy instruments for implementation, with variations between regions. The Italian regions typically include regulatory, economic/financial and/or communication policy instruments. On the other hand, the Croatian counties foresee the use of regulatory and/or voluntary policy instruments for implementation. These variations in the use of policy instruments underline the different approaches taken by these countries to implement their respective measures.

The strategies involve a multitude of stakeholders from public and private spheres (Fig. 7.6), showing diverse collaborations among public authorities of different levels, research institutions, societal groups, private entities, and citizens, favoring mostly the involvement of public authorities followed by non-governmental organizations (NGOs), business associations, services, and voluntary associations. Overall, despite

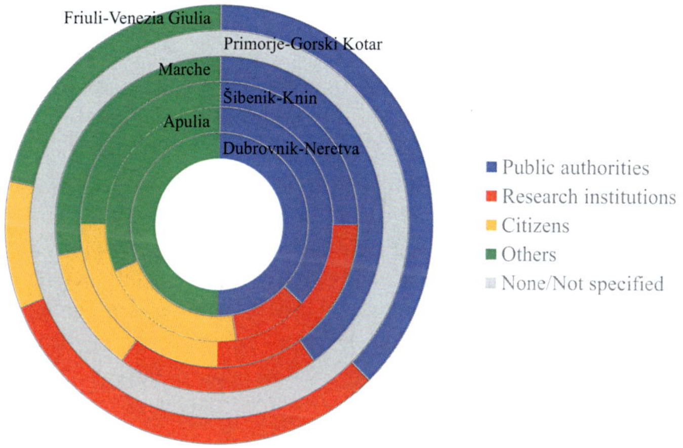

Fig. 7.6 Proportion of stakeholders involved in the development of the strategies by region. The category "others" includes non-governmental organizations (NGOs), private institutions, services, voluntary associations, etc.

extensive collaboration endeavors, communication often leans towards top-down, suggesting room for more inclusive decision-making.

Throughout these regions, while there are extensive collaborative endeavors, communication styles often lean toward top-down models, signaling potential opportunities for greater inclusivity and engagement in decision-making processes.

Strategies are often integrated into new or existing policies and results are always evaluated through monitoring. In Friuli-Venezia Giulia, most mitigation strategies are integrated into existing policies, with ongoing progress and evaluation based on monitoring and objective criteria, which is well aligned with EU legislation. In Primorsko-Goranska County, the strategy is integrated into policies and faces budgetary challenges, but benefits from strong planning links. Marche Region strategies, mostly integrated into policies, use various funding sources and are evaluated through monitoring, highlighting various strengths such as technology use and waste reduction, but facing weaknesses such as employment opportunities and funding limitations. Šibenik-Knin County emphasizes the linkages between relevant sectoral plans as a strength, but struggles with budgetary constraints for adaptation strategies, emphasizing the interdisciplinary nature of planning. Apulia Region strategies, mostly integrated into policies, are being monitored for evaluation, highlighting strengths in prevention and technology, but facing challenges in time and data availability. Dubrovnik-Neretva County relies on European funding and monitoring, but lacks future climate scenarios, which is a significant weakness.

Finally, in terms of lessons learned, in Italy, the Friuli-Venezia Giulia Region advocates cross-border cooperation to promote interaction between different stakeholders, taking into account not only economic benefits but also environmental, energy and social aspects in decision-making. Similarly, the Marche Region emphasizes the essence of cooperation between political, non-political and scientific entities, relying on meticulous planning and real-time monitoring to compensate for the lack of previous experience. On the other hand, the Apulia Region aims to increase citizens' awareness of risks while striving for a sustainable quality of life and economic prosperity in the region. In Croatia, the County of Primorje-Gorski Kotar advocates the targeted allocation of funds, the County of Šibenik-Knin stresses the need for a holistic and integrated approach, and the County of Dubrovnik-Neretva emphasizes the importance of interdisciplinary planning.

In summary, the regional strategies highlight the need to emphasize cross-border cooperation, careful planning, increased citizen awareness, targeted resource allocation, holistic approaches, and interdisciplinary planning as critical aspects of effective climate change strategies. These lessons emphasize the need for comprehensive, inclusive, and adaptive approaches to address the multiple challenges posed by climate change across regions.

References

Alpenallianz (2017) Carta di Budoia per l'azione dei Comuni alpini nell'adattamento locale ai cambiamenti climatici [Budoia Charter for the action of Alpine Municipalities in local adaptation to climate change]. https://www.legambiente.it/wp-content/uploads/2023/06/Carta-di-Bua.pdf

Apulia Region (2006) Regolamento Regionale 22 agosto 2006, n. 13. Misure urgenti per il contenimento dell'inquinamento luminoso e per il risparmio energetico [Urgent measures to reduce light pollution and save energy]. https://bussolanormativa.consiglio.puglia.it/public/leges/LeggeNavscroll.aspx?id=11607

Apulia Region (2008a) Regolamento per aiuti agli investimenti delle Piccole e Medie Imprese nel risparmio energetico, nella cogenerazione ad alto rendimento e per l'impiego di fonti di energia rinnovabile in esenzione ai sensi del Regolamento (CE) N. 800/2008 [Aid for investments of small and medium enterprises in energy saving, in high-efficiency cogeneration and in the use of renewable energy sources]. http://www5.consiglio.puglia.it/Giss8/8SagArchivio.nsf/(InLinea)/provv.amm.-539-VIII/$File/delg1885.pdf?OpenElement

Apulia Region (2008b) Legge Regionale 10 giugno 2008, n. 13 "Norme per l'abitare sostenibile" [Standards for sustainable living]. https://www.anit.it/wp-content/uploads/2015/03/LR132.pdf

Apulia Region (2015a) Programma Operativo Regionale (POR) Puglia 2014–2020 [Regional operational programme 2014–2020]. https://por.regione.puglia.it/en/por

Apulia Region (2015b) Programma di Sviluppo Rurale (PSR) Puglia 2014–2022. Programma di sviluppo rurale Regione Marche 2014–2022 [Regional rural development programme 2014–2022]. https://psr.regione.puglia.it/en/il-programma

Apulia Region (2017) Recupero e riutilizzo di eccedenze, sprechi alimentari e prodotti farmaceutici [Recovery and reuse of surplus, food waste and pharmaceutical products]. https://burp.regione.puglia.it/documents/20135/1049418/Bollettino+numero+58+-+Ordinario+-+anno+2017.pdf/19b1d434-e3b5-0cb7-2e04-5dcf4d5e7b56?t=1622804552494

Croatian Air Protection Act (2011) Official gazette #130/11, 47/14, 61/17. http://narodnenovine.nn.hr/clanci/sluzbeni/2011_11_130_2601.html

Croatian Ministry of Environmental Protection and Energy (2021) Low carbon development strategy of the Republic of Croatia until 2030 with a view to 2050. https://mingor.gov.hr/UserDocsImages/klimatske_aktivnosti/odrzivi_razvoj/NUS/lts_nus_eng.pdf

Dubrovnik-Neretva County (2018) Program zaštite okoliša Dubrovačko-Neretvanske županije za razdoblje 2018–2021 [Environmental protection program for Dubrovnik-Neretva County for the period 2018–2021]. https://www.zzpudnz.hr/Portals/0/literatura/Tekst_PZO_DNZ_Final_4.pdf?ver=B8sYAO6Ao7Mcl84xyyoirw%3d%3d

European Commission (2011) Roadmap for moving to a competitive low-carbon economy in 2050. COM(2011) 112 final. https://eur-lex.europa.eu/LexUriServ/LexUriServ.do?uri=COM:2011:0112:FIN:EN:PDF

European Commission LIFE TRUST (n.d.) Tool for regional—scale assessment of groundwater storage improvement in adaptation to climate change (TRUST). https://webgate.ec.europa.eu/life/publicWebsite/project/LIFE07-ENV-IT-000475/tool-for-regional-scale-assessment-of-groundwater-storage-improvement-in-adaptation-to-climate-change-trust

European Union Horizon 2020 NOEMIX (n.d.) NOEMIX—new mobility in Friuli-Venezia Giulia. https://www.noemix.eu/en/

Friuli-Venezia Giulia Agency for Environmental Protection (2018) Studio conoscitivo dei cambiamenti climatici e di alcuni loro impatti in Friuli-Venezia Giulia [Cognitive study of climate change and some of their impact in Friuli-Venezia Giulia]. https://www.meteo.fvg.it/clima/clima_fvg/03_cambiamenti_climatici/01_REPORT_cambiamenti_climatici_e_impatti_per_il_FVG/impattiCCinFVG_marzo2018.pdf

Friuli-Venezia Giulia Autonomous Region (2007) Piano regionale di miglioramento della qualità dell'aria [Regional plan for the improvement of the air quality]. https://www.regione.fvg.it/rafvg/export/sites/default/RAFVG/ambiente-territorio/allegati/DGR913ALL1.pdf

Friuli-Venezia Giulia Autonomous Region (2011) Piano regionale delle infrastrutture di trasporto, della mobilità delle merci e della logistica [Regional plan of transport infrastructure, freight mobility and logistic]. https://www.regione.fvg.it/rafvg/cms/RAFVG/infrastrutture-lavori-pubblici/infrastrutture-logistica-trasporti/FOGLIA18/articolo.html

Friuli-Venezia Giulia Autonomous Region (2013) Politica regionale di coesione 2014–2020. Principali criticità ambientali e azioni di risposta per il territorio della Regione Friuli-Venezia Giulia [Main environmental criticalities and response actions for the territory of the Friuli-Venezia Giulia Region]. https://www.regione.fvg.it/rafvg/export/sites/default/RAFVG/ambiente-territorio/tutela-ambiente-gestione-risorse-naturali/FOGLIA209/allegati/Rapporto_criticitx_allegato_a_DGR_2405.pdf

Friuli-Venezia Giulia Autonomous Region (2015a) Impatti dei cambiamenti climatici sul territorio fisico regionale [Impacts of climate change on the regional physical territory]. https://www.regione.fvg.it/rafvg/export/sites/default/RAFVG/ambiente-territorio/geologia/FOGLIA22/allegati/Impatti_dei_cambiamenti_climatici_sul_territorio_fisico_regionale.pdf

Friuli-Venezia Giulia Autonomous Region (2015b) Piano energetico regionale [Regional energy plan]. https://www.regione.fvg.it/rafvg/cms/RAFVG/ambiente-territorio/energia/FOGLIA111/#id3

Friuli-Venezia Giulia Autonomous Region (2015c) Programma di Sviluppo Rurale (PSR) 2014–2020 [Rural development program]. https://www.regione.fvg.it/rafvg/cms/RAFVG/economia-imprese/agricoltura-foreste/psr-programma-sviluppo rurale/FOGLIA119/

Friuli-Venezia Giulia Autonomous Region (2017) Piano di Assetto Idrogeologico dei bacini Regionali (PAIR) [Plan for the geo-hydrological management for of the regional basins]. https://www.regione.fvg.it/rafvg/cms/RAFVG/ambiente-territorio/geologia/FOGLIA24/

Friuli-Venezia Giulia Autonomous Region (2018) Piano d'azione della Regione Friuli-Venezia Giulia per gli acquisti verdi [Friuli-Venezia Giulia Region action plan for green purchases]. https://mtom.regione.fvg.it/storage/2018_367/Allegato%201%20alla%20Delibera%20367-2018.pdf

Intergovernmental Panel on Climate Change—IPCC (2013) Climate change 2013: the physical science basis. Stockholm

Interreg Greece-Italy OFIDIA (n.d.) Operational Fire Danger Prevention Platform (OFIDIA). https://ec.europa.eu/regional_policy/it/projects/italy/a-new-monitoring-system-to-prevent-wildfires

Italian Ministry of Economic Development (2012) Decreto Ministeriale 15/03/2012. Definizione e qualificazione degli obiettivi regionali in materia di fonti rinnovabili e definizione della modalità di gestione dei casi di mancato raggiungimento degli obiettivi da parte delle regioni e delle provincie autonome (c.d. burden sharing). http://www.energia.provincia.tn.it/binary/pat_agenzia_energia/normativa/D.M._15_3_2012.1334145125.pdf

Marche Region (2005) Piano Forestale Regionale [Regional forest plan]. https://www.consiglio.marche.it/banche_dati_e_documentazione/iter_degli_atti/paa/pdf/d_am106_8.pdf

Marche Region (2007) Strategia Regionale d'Azione ambientale per la Sostenibilità (STRAS 2006–2010) [Regional environmental action strategy for sustainability 2006–2010]. https://ambiente.regione.marche.it/Portals/0/Ambiente/STRAS/Stras_pubblicata.pdf

Marche Region (2011) Tavolo regionale per le foreste [Regional forest table]. https://www.ciamarche.org/bancadati/legge-forestale-regionale

Marche Region (2014) Piano Regolatore degli Acquedotti della Regione Marche [Regulator plan for aqueducts]. https://www.regione.marche.it/Regione-Utile/Ambiente/Tutela-delle-acque/PRA

Marche Region (2015a) Piano Regionale di Gestione dei Rifiuti (PRGR) [Regional waste management plan]. http://www.regione.marche.it/Regione-Utile/Ambiente/Rifiuti-e-inquinamento/Rifiuti#Pianificazione

Marche Region (2015b) Programma di sviluppo rurale Regione Marche 2014–2022 [Regional rural development plan 2014–2022]. https://www.regione.marche.it/Entra-in-Regione/Psr-Marche/Psr-2014-2020/Cos%C3%A8-il-PSR

Marche Region (2016) Piano Energetico Ambientale Regionale (PEAR) [Regional environmental energy plan]. https://www.regione.marche.it/Regione-Utile/Energia/Piano-Energetico-Ambientale-Regionale

Marche Region (2017b) Piano regionale delle attività di previsione, prevenzione e lotta attiva contro gli incendi boschivi 2017–2019 [Regional plan of forecasting, prevention and active fight against forest fires 2017–2019]. http://www.regione.marche.it/Portals/0/Agricoltura/Foreste/Normativa_Regionale/DGR%20792_2017%20Piano%20Antincendi%20Boschivi.pdf

Marche Region (2017a) Piano straordinario per lo sviluppo, la promozione e la valorizzazione delle Marche. Integrazione alla deliberazione 13/2015. X Legislatura [Extraordinary plan for the development, promotion and valorization of the Marche Region]. https://www.consiglio.marche.it/banche_dati_e_documentazione/iter_degli_atti/paa/pdf/d_am41_10.pdf

Marche Region (2019) Piano di Gestione Integrata delle Zone Costiere (Piano GIZC) [Integrated coastal zone management plan 2019]. http://www.regione.marche.it/Regione-Utile/Paesaggio-Territorio-Urbanistica-Genio-Civile/Difesa-della-costa#Piano-GIZC-2019

Primorje-Gorski Kotar County (2019) Program zaštite zraka, ozonskog sloja, ublažavanja klimatskih promjena i prilagodbe klimatskih promjenama u Primorsko-Goranska županiji za razdoblje 2019–2022 [Program for air protection, ozone layer protection, climate change mitigation and adaptation to climate changes]. https://www2.pgz.hr/doc/dokumenti/savjetovanje-s-javnoscu/2019/zrak/Nacrt%20programa.pdf

Priority Actions Programme Regional Activity Centre (2016) Integrated coastal zone management plan of the Šibenik-Knin County. https://maritime-spatial-planning.ec.europa.eu/practices/coastal-plan-sibenik-knin-county

United Nations Development Programme—UNDP (2008) Human development report for Croatia: a climate for change: climate change and its impacts on society and economy in Croatia. UNDP, Zagreb, Croatia

Chapter 8
Aligning Regional Strategies with Climate Trends

In 2017, the European Environment Agency presented a largely indicator-based assessment of past and future climate change, its impacts and the associated vulnerabilities and risks to ecosystems, human health, and society in Europe, based on a wide range of observations and model simulations. They identified biogeographical regions particularly vulnerable to the impacts of climate change. The report shows a list of observed and projected climate change impacts that affect the main biogeographical regions in Europe, namely Arctic region, Atlantic region, Mountain regions, Coastal zones and regional seas, Boreal region, Continental region, and Mediterranean region. The regionalization is based on the map of biogeographic regions produced by the Convention on the Conservation of European Wildlife and Natural Habitats (Bern Convention), with some adjustments due to limited data availability for some regions. Based on this classification, the Adriatic area encompasses the following biogeographical zones: (a) Mediterranean region, (b) Coastal zones and regional seas, and (c) Mountainous regions. According to the report on climate change, impacts and vulnerability in Europe (EEA 2017), the Mediterranean region is facing decreasing water availability, crop yields, and biodiversity, as well as increasing droughts, forest fires and negative impacts on human health and livestock due to decreasing precipitation and increasing temperatures, especially in summer. Coastal areas will face increased flooding due to sea level rise and storm surges, and changes in marine ecosystems due to ocean warming and acidification, affecting fisheries and water quality. Mountains will experience loss of snow and glaciers and shifts in biodiversity due to temperature increases and changes in precipitation patterns, affecting water resources and tourism, while floods and landslides will become more frequent and severe.

The activities carried out in the framework of the Interreg Italy-Croatia RESPONSe project have allowed to downscale the impacts of climate change on the Adriatic Basin, with specific focus on the northern, central, and southern areas. The results of such climate analysis are reported in the "Deliverable 3.1 Climate change assessment: observational base study" (2020a) and in the "Deliverable 3.2.1 Climate

E. Gioia, *Climate Change Adaptation in the Adriatic Coastal Areas*, Disaster Risk Reduction, https://doi.org/10.1007/978-981-97-6655-0_8

Table 8.1 Climate change variables (sea, precipitation, wind, and air temperature) and related indices considered in the regional mitigation and adaptation strategies

Sea	Precipitation	Wind	Air temperature
Sea level rise	Average annual precipitation	Wind seasonality	Average annual temperature
Sea temperature	Average seasonal precipitation	Interaction wind and sea	Urban heat island effect
Salinity increase	Number of storms	Wind intensity	
Coastal erosion and transversal displacements	Drying wetlands		
Number of storm surges			
Coastal floods			
Surface waves			

change projections assessment: analysis of regional climate models' simulations" (2020b), available for public viewing at the RESPONSe project website: https://programming14-20.italy-croatia.eu/web/response. Part of these analyses then merged into the publications of Sokol Jurković et al. (2022) and Tojčić et al. (2023).

The purpose of this section is to compare the local climate assessment of observed and projected changes with the climate scenarios considered to inform the analyzed regional strategies (Chap. 7), in order to understand if the path taken by the decision makers of the Adriatic region is compatible with the indications of the climate scientists. Indeed, most of the regional strategies analyzed relied on climate scenarios and projections to inform that strategy. The specific climate variables considered in the development of the regional strategies are (i) sea, (ii) precipitation, (iii) wind, and (iv) air temperature, and the related indices are listed in Table 8.1.

8.1 Observed Impacts of Climate Change

Similarly to the strategies, the variables considered in the analysis of the observed impacts of climate change in the Adriatic Basin are: (i) sea, (ii) precipitation, (iii) wind, and (iv) air temperature (Table 8.2). Each variable has been listed in different indices over different time periods. Specifically, data on sea observations (i.e., sea level height, sea temperature at different depths, salinity, sea surface temperature trends), precipitation patterns (i.e., heavy precipitation and consecutive wet or dry days), wind characteristics (i.e., mean, gust, and maximum speed), and air temperature variations (i.e., maximum and minimum daily temperatures, summer days, and tropical nights) are considered. Data range from annual to decades or centuries.

The results of these analyses were compared with those actually used in the development of the regional mitigation and adaptation strategies in northern, central

Table 8.2 Climate change variables (sea, precipitation, wind, and air temperature) and related indices considered for the climate observation analysis

Sea	Precipitation	Wind	Air temperature
Sea level height for the periods 1955–2014 [mm/year]	Annual number of wet days (days when total daily precipitation RR ≥ 1 mm) [N/decade]	Mean annual wind speed (m/s/century)	Mean annual daily maximum temperature (TX) [°C/decade]
Sea temperature averaged for depths between 30 and 100 m for the period 1963–2016 [°C/century]	Annual number of heavy precipitation days (days when total daily precipitation RR ≥ 20 mm) [N/decade]	Maximum annual wind speed (m/s/century)	Mean annual daily minimum temperature (TN) [°C/decade]
Salinity (practical salinity unit) averaged for depths between 30 and 100 m for the period 1963–2016 [PSU/century]	Annual number of consecutive wet days (CWD) periods (period of at least 5 days with total daily precipitation RR ≥ 1 mm) [N/decade]	Mean annual wind gust (m/s/century)	Annual number of summer days (days when TX > 25 °C) [N/decade]
Sea temperature averaged for depths between 30 and 100 m for the period 1963–2016 [°C/century]	Annual number of heavy precipitation days (days when total daily precipitation RR ≥ 20 mm) [N/decade]		Mean annual daily minimum temperature (TN) [°C/decade]
Salinity (practical salinity unit) averaged for depths between 30 and 100 m for the period 1963–2016 [PSU/century]	Annual number of consecutive wet days (CWD) periods (period of at least 5 days with total daily precipitation RR ≥ 1 mm) [N/decade]		Annual number of summer days (days when TX > 25 °C) [N/decade]
Sea surface temperature linear trends (SST) for the period 1964–1978 [°C/decade]	Time series of the total annual precipitation amounts (RR) [mm/day/decade]		Annual number of tropical nights (days when TN < 20 °C) [N/decade]
	Annual number of consecutive dry days (CDD) periods (period of at least 5 days with total daily precipitation RR ≤ 1 mm) [N/decade]		

and southern Adriatic. This comparison can be used to assess how well the indices chosen in the strategies match those evaluated in the climate analysis.

As for the sea, the climate analysis showed an upward trend for the sea level measured in both the periods 1955–2014 and 1990–2014, but in the period 1990–2014 the values of mm/year are greater than the first period by about four times. The height of the sea level increases from the north to the south of the Adriatic. On the other hand, the regional strategies considering the increase of the sea level were concentrated in the north and in the center. In the southern Adriatic, only the Croatian side considered such an index.

The average salinity for depths between 30 and 100 m has been analyzed only in the station Hvar (center, Croatia) for the period 1963–2016 and 1993–2016. They both have a positive trend, higher in the second period. In fact, only the Šibenik-Knin (central) and the Friuli-Venezia Giulia (northern) regions considered this index.

The sea surface temperature in the northern and central Adriatic shows a decreasing trend for the period 1964–1978 and an increasing trend for the period 1979–2015. In this case, the downward trend is lower in the northern stations and the upward trend is higher in the central stations. The mean sea temperature for depths between 30 and 100 m, analyzed only for the Hvar station (central, Croatia), shows upward trends in the period 1963–2016, with much higher temperature increases in the last period 1993–2016. The regional strategies that considered the increase of sea temperature were only in the regions of Friuli-Venezia Giulia (north) and Marche (center).

The regional strategies also considered: increase of coastal flooding (Marche and Šibenik-Knin, center; Apulia, south); increase of storm surges (Friuli-Venezia Giulia, north, and Šibenik-Knin, center); increase of coastal erosion and sediment transport (Marche, center; Apulia, south).

Concerning the variable precipitation, the results of the climatic analysis showed that the variation of the related indices is not homogeneous among all regions. In particular, the index of the annual number of wet days shows a decreasing trend in the Italian regions and an increasing trend in the Croatian regions, although not always significant. In addition, the annual number of consecutive wet days index shows an increasing trend for Friuli-Venezia Giulia (north, Italy), a decreasing trend for Primorje-Gorski Kotar (north, Croatia) and no trend for all other regions. Furthermore, the annual number of heavy precipitation days shows a general increasing trend for all regions, with the highest values for the Croatian regions of Primorje-Gorski Kotar (north) and Šibenik Knin (center). Consistently, the regional strategies of Friuli-Venezia Giulia (north), Marche and Šibenik-Knin (center), and Dubrovnik-Neretva (south) take into account the increasing number of heavy precipitation days.

Time series of total annual precipitation do not show trends for all project areas, and most values are not significant. On the other hand, the strategies of Primorje-Gorski Kotar (north), Dubrovnik-Neretva and Apulia (south) consider a decrease of annual precipitation.

Finally, the analyzed annual number of consecutive dry days shows no statistically significant trends, but the increase of dry periods and droughts has been considered in the strategies of almost all the regions, with the exception of the Dubrovnik-Neretva County (south).

Regarding the variable wind, the climatic analysis showed the increase of the mean annual wind speed and wind gusts for the locations in Friuli-Venezia Giulia (north), Šibenik-Knin (center), Apulia and Dubrovnik-Neretva (south); the increase of both annual mean and maximum wind speed for Primorje-Gorski Kotar (north) and Marche (center); the decrease of the maximum annual wind speed and wind gusts for the locations of Friuli-Venezia Giulia (north), Šibenik-Knin (center), Apulia and Dubrovnik-Neretva (south); the decrease of both annual mean and maximum of the wind gusts for Primorje-Gorski Kotar (north) and Marche (center).

The regional strategies of Friuli-Venezia Giulia (north), Šibenik-Knin (center), Apulia and Dubrovnik-Neretva (south) considered the wind variable but in terms of seasonality, wind-sea interaction and intensity.

As for the variable air temperature, the climate analysis showed that the air temperature related indices have increasing and statistically significant trends in all regions. In particular, the increase of the mean annual daily maximum temperature is higher in the north and in the center, while the mean annual daily minimum temperature is higher in the central Adriatic region. Moreover, the analysis shows that in the north the annual number of summer days has the highest increase, while the annual number of tropical nights has the lowest increase.

All the regions considered in their strategies the increase of the average air temperature and in the Apulia (south) region the strategies also focused on the heat island effects.

8.2 Projected Scenarios of Climate Change

Similar to the observations, the variables considered in the analysis of the projected impacts of climate change in the Adriatic Basin are: (i) sea, (ii) precipitation, (iii) wind, and (iv) air temperature (Table 8.3). Each variable has been listed in different indices. Specifically, data on sea observations (i.e., sea surface temperature, salinity and height), precipitation patterns (i.e., mean and median daily precipitation), wind characteristics (i.e., daily wind speed) and air temperature variations (i.e., mean and median daily temperatures, heating, and cooling degree days) are considered. In this case, a fifth analysis on the future impact of climate change on tourism activities has also been added.

Regarding the variables describing the sea, the Sea Surface Temperature (SST) and Sea Surface Salinity (SSS) indices in the climate analysis show increasing trends, as also emerged from the analysis of some regional strategies of Friuli-Venezia Giulia. Both indices will increase more in the medium future (2041–2070) than in the near future (2011–2040) and this is coherent with the rest of Europe. The analysis of the Sea Surface Height (SSH) index shows a decrease in the near future in the whole Adriatic Sea and a subsequent increase in the medium future. However, the range of possible changes in SSH is too conservative compared to similar studies.

Table 8.3 Climate change variables (sea, precipitation, wind, air temperature, and tourism) and related indices considered for the climate projection analysis

Sea	Precipitation	Wind	Air temperature	Tourism
Sea surface temperature (SST)	Mean daily precipitation	Daily wind speed	Mean daily temperature	Climate Index for Tourism (CIT)
Sea surface salinity (SSS)	Median daily precipitation		Median daily temperature	
Sea surface height (SSH)			Heating degree day (HDD)	
			Cooling degree day (CDD)	

Consideration of the additional coupled regional atmosphere–ocean–sea models is recommended as soon as their results become publicly available.

Sea Surface Height (SSH) is the only index considered in some of the strategies analyzed (Friuli-Venezia Giulia and Primorje-Gorski Kotar, north) and only for the case of sea level rise. Šibenik-Knin County (central) evaluates the SSH to check if it is increasing. Other regions do not mention studies on these indices, even though the changing state of the sea will affect many economic activities and will strongly change the shape of the Adriatic basin. This consequence could also change the demographic and urban characteristics of the area.

As for the variable precipitation, the climate analysis considered the median and mean ensemble values of mean seasonal precipitation related to RCP4.5 (moderate emissions scenario) and RCP8.5 (high emissions scenario) climate scenarios (IPPC 2001). The results of the studies indicate that in the near (2011–2040) and medium (2041–2070) future the seasonal precipitation will change in the whole Adriatic region. In particular, in the RCP4.5, scenario precipitation will increase in all seasons except June–July–August (JJA) in the near future; precipitation will increase in September–October–November (SON) and will decrease in JJA in the medium future. In the RCP8.5 scenario, precipitation will decrease in JJA in both 2011–2040 and 2041–2070, besides a moderate increase in December–January–February (DJF) in the near future.

These studies for future precipitation projections have also been carried out by some regions such as Friuli-Venezia Giulia and Primorje-Gorski Kotar (north) and Dubrovnik-Neretva (south). However, in the strategies they declared a general decrease in the amount of precipitation in the entire year, not a variation in the seasonal distribution. Nevertheless, the seasonal analysis is important for all regions because it will describe the future consequences of the change in precipitation distribution, such as dry periods or sudden storms in other seasons.

Concerning the wind variable, the climate analysis examined the daily wind speed in the RCPs scenarios, and the results establish that the changes in wind speeds by the end of the twenty-first century over the Adriatic region are small.

Accordingly, the wind variable is considered only in the County of Primorje-Gorski Kotar (north), where it is reported that the mean wind speed in summer and fall will increase for the period 2011–2040 and will remain unchanged for the period 2041–2070, while the maximum speed will remain unchanged for the period 2011–2040 and will decrease for the period 2041–2070. The consideration of the variable wind is important for all regions, because the consequences could have serious impacts on sea waves or important innovations, i.e., in wind production.

As for the variable air temperature, the climate analysis examined the current relative mean change with the future relative mean change, in the Adriatic region, for each season and in relation to the RCPs scenarios in the two periods 2011–2040 and 2041–2070. For the RCP4.5 scenario, in the first period, the near future, the temperature will increase in all seasons. In the intermediate future, the lowest increase will occur in March–April–May (MAM), while the highest increase will occur in JJA. The range of increase is between 1–1.5 and 2–2.5 °C. For RCP8.5 scenario, the temperature increase is lowest than in RCP4.5 for the period 2011–2040 and it is much higher during the period 2041–2070. In the near future, the range of temperature increase is between 0.5 and 1.5 °C, with a minimum in MAM and a maximum in JJA. During the middle future, the range will increase between 2 and 3 °C from MAM to JJA.

The variable air temperature studies also included the evaluation of two indices, Heating Degree Day (HDD) and Cooling Degree Day (CDD), in relation to the RCP4.5 and RCP8.5 scenarios. These two indices are commonly used to derive average air temperature data. The HDD quantifies the energy required to heat a building when the average daily temperature is less than 18 °C. The CDD measures the energy required to cool buildings when the average daily temperature is above 18 °C. The analysis shows that the HDD will decrease in the near and medium future in southern Europe and in the Mediterranean regions, which means that the average hot days will increase; on the contrary, the CDD will increase and consequently the cold days will decrease during the year. In summary, the air temperature will increase in future and the number of cold days will decrease.

These indices are also considered in some strategies of the Friuli-Venezia Giulia Region and the Primorje-Gorski Kotar County (north) and in the Dubrovnik-Neretva County (south). The other regions do not consider these indices in their strategies. However, future increases in air temperature and related consequences, such as desertification and reduced energy consumption for heating, coupled with progressively higher energy demand for cooling indoor environments, should necessarily be taken into account.

In regard to the variable tourism, the Climate Index for Tourism (CIT) is an index used for Mediterranean climate change studies that allows the assessment of the weather resource sun, sea and sand (3S) (Amengual et al. 2011). Tourism is one of the largest European Mediterranean and Adriatic economic sectors worldwide, and beach-based vacations are an important source of resource income (Amengual et al. 2011). 3S tourism is strongly linked to a number of atmospheric variables such as temperature, precipitation, relative humidity, cloudiness and wind speed (De Freitas 1985). Global environmental change threatens these conditions, and 3S tourism transportation activities directly affect climate change through greenhouse gas emissions and the seasonality of the industry in many regions of the world (Amengual et al. 2011).

Tourism activities depend on weather conditions, and in the Adriatic Sea the climate analysis evaluated a decrease in the number of days with ideal conditions in summer (− 30 days), but an increase in the number of days with acceptable conditions in spring (+ 30 days) and ideal conditions in autumn. This could mean a change in the seasonality of tourism activities caused by the effect of unacceptable, acceptable and ideal changing conditions. According to the analysis, unacceptable conditions will shift in future from November to March at 7 a.m., acceptable conditions from April to October at 7 p.m., and ideal conditions from May to September at 7 a.m., increasing from the northern to the southern Adriatic, and from April to October at 2 p.m. Overall, the climatic preference would extend from May to October instead of June, July and August.

The 3S tourism activities considered for the Adriatic Basin include common sports activities that are played in the Basin. The expected changes in the activities are as follows:

- *Cycling* will become unsuitable in summer, more suitable in spring and fall, and there are no significant changes in winter.
- *Cultural activities and soccer* will not change their comfortable condition.
- *Tourist potential* would deteriorate in summer.
- *Golf* will decrease in winter, will increase slightly in summer and will remain the most suitable in spring and fall.
- *Motor boating* will increase in all seasons.
- *Sailing* will slightly decrease in summer and the comfort period will not change (spring and fall).
- *Hiking* will decrease the ideal condition in summer.

Tourism is one of the vulnerable sectors and topics included in the strategies of all analyzed Italian regions and Croatian counties, but no strategy evaluates future projections of this sector. For this reason, a good assessment of the future state of tourism could be a necessity for future or updated strategies. Knowing the future projections and with a different planning, it could be possible to reduce the vulnerability of tourism by reorganizing the program of touristic activities, or the related activities, i.e., holidays, opening of touristic facilities, exhibitions, tours.

8.3 Summary

The information on the challenges addressed by the analyzed regional strategies was compared with the impacts of climate change on sea level, precipitation, wind, and air temperature in the Adriatic Basin. The aim was to understand if the implemented strategies are consistent with the actual impacts of climate change observed and expected in these regions. In fact, all changes due to climate change influence current and future human activities, so adaptation and mitigation strategies must take them into account during planning.

Although not every strategy is based on robust climate scenarios and projections, some of the climate indices considered are similar. However, the list of possible indices is not exhaustive, and additional indices may be considered when preparing the next generation of climate change adaptation strategies.

The comparison between the climate analysis and the regional strategies in the Adriatic regions shows some similarities and discrepancies. Sea level rise, salinity and sea temperature trends are consistent in the analysis but are mainly addressed in the northern and central regional strategies, overlooking the southern Adriatic. Few regions address the implications of sea surface height (SSH), which is expected to increase in the medium term, highlighting the need for a broader understanding given its impact on the characteristics of the Adriatic.

Precipitation seasonal trends diverge, decreasing in Italian regions and increasing in Croatian areas, information that should shape the strategies accordingly. For example, the number of heavy precipitation days increases in all regions and is reflected in the strategies, while the expected dry periods in the strategies contrast with unclear trends in the climate analysis, except in the Dubrovnik-Neretva County. Precipitation projections, which vary between scenarios, are acknowledged in some strategies, but most emphasize a general decrease without addressing critical seasonal shifts.

Wind speed changes are minimal, with Primorje-Gorski Kotar reflecting on their implications, particularly for sea waves and wind energy.

Air temperature projections, which show increasing trends and significant index changes, are considered in a few strategies, such as Apulia, highlighting the importance of considering future temperature increases and related consequences.

Finally, the climate analyses have included the impact of weather on tourism, a sector that lacks future projections in the strategies, indicating the need for anticipatory planning to reduce vulnerability in this sector.

Such a comparison underlines the need for a more consistent integration of climate trends into strategies throughout the Adriatic area.

References

Amengual A, Homar V, Romero R, Alonso S, Ramis C (2011) Projections of the climate potential for tourism at local scales: application to Platja de Palma, Spain. Royal Meteorological Society (RMetS). https://doi.org/10.1002/joc.2420

De Freitas CR (1985) Assessment of human bioclimate based on thermal response. Int J Biometeorol 29:97–119

European Environment Agency—EEA (2017) Climate change, impacts and vulnerability in Europe 2016. EEA report no. 1/2017, Publications Office of the European Union, Luxembourg. https://www.eea.europa.eu/publications/climate-change-impacts-and-vulnerability-2016

Interreg Italy-Croatia RESPONSe (2020a) Deliverable 3.1: climate change assessment: observational base study. https://programming14-20.italy-croatia.eu/web/response/docs-and-tools#docsTools_3129392

Interreg Italy-Croatia RESPONSe (2020b) Deliverable 3.2.1: climate change projections assessment: analysis of regional climate models' simulations. https://programming14-20.italy-croatia.eu/web/response/docs-and-tools#docsTools_3129392

Sokol Jurković R, Güttler I, Pasarić Z (2022) Bivariate bias correction of the regional climate model ensemble over the Adriatic region. Int J Climatol 42(11):5826–5847. https://doi.org/10.1002/joc.7564

Tojčić I, Denamiel C, Vilibić I (2023) Kilometer-scale trends and variability of the Adriatic present climate (1987–2017). Clim Dyn 61:2521–2545. https://doi.org/10.1007/s00382-023-06700-2

Intergovernmental Panel on Climate Change - IPCC (2001), Climate Change 2001: Impacts, Adaptation and Vulnerability, Cambridge University Press, Cambridge.

Chapter 9
From Strategy to Action: An Open Database for Climate Adaptation

Adaptation strategies are typically shaped by regulations at various levels (supra-national, national and regional). However, the effectiveness of generic options is often limited by the specific local characteristics of each location. As a result, public authorities could benefit from a tool that facilitates the translation of strategies into concrete local actions. In this context, the Climate Menu toolkit is an innovative online compendium of adaptation measures designed to catalyze responses to climate change. Moreover, some of these actions have been selected and tailored to the unique characteristics of the Adriatic study areas.

This work is mainly addressed to public authorities as a comprehensive, easy-to-use tool covering adaptation actions, grouped by and adapted to the characteristics of the Adriatic study areas. Such a unique review will serve as a database of actions to be selected to implement an adaptation strategy in a specific area. Public authorities can choose from these state-of-the-art measures to initiate adaptation planning at the local level. The resulting database, labeled the Climate Menu for Adriatic Regions toolkit, is accessible as a web platform.

9.1 Analysis Criteria

The methodology employed in this work was guided by three primary criteria.

Initially, the analysis centered on regional strategies within the Adriatic Basin, acknowledging their integration into broader European strategies through a hierarchical management system. This phase marked a pivotal transition from examining overarching strategies to focusing on measures and actions. The shift aimed to delve into specific problems and practices already implemented in the Adriatic regions.

Subsequently, a comprehensive integration of additional actions was undertaken, drawing from policies and initiatives implemented within Europe and also beyond its borders. This approach sought to construct a holistic understanding of effective

E. Gioia, *Climate Change Adaptation in the Adriatic Coastal Areas*, Disaster Risk Reduction, https://doi.org/10.1007/978-981-97-6655-0_9

climate strategies relevant to the Adriatic. Emphasizing the adoption of established best practices facilitated the identification and evaluation of successful initiatives.

Lastly, to prioritize adaptation efforts based on the actual impacts of climate change, a meticulous selection of actions capable of addressing the present and future climate trends in the Adriatic macro-areas was carried out. This step aimed to align efforts with the real and projected effects of climate change in the region.

9.2 Data and Metadata Structure

The Climate Menu toolkit includes a list of state-of-the-art adaptation actions that address the specific climate change challenges of the Adriatic region. These actions define objectives, actors, expected results and timelines for the implementation of adaptation planning, taking into account local needs. Consideration has also been given to existing good practices.

The actions have been structured in individual technical sheets, which are organized into two main contents:

- the data (Table 9.1), that is the main information related to the action (i.e., title, objective, description, expected results, results indicators, involved actors, expected timeline for the action, good practices, critical issues);
- the metadata (Table 9.2), that is the descriptive information that provides additional knowledge about the data and is essential to browse for specific actions through online filters (i.e., scope, type, sector, climate impact, scale of implementation, date of last update, source).

Table 9.1 List of the "data" information included in the technical sheets to describe the action

Data	Content
Title	Title of the action
Objective	Purpose of the action
Description	Description of the planned or adopted action and implementation methods
Expected results	Benefits caused by the implementation of the action
Results indicators	Quantitative or qualitative variables that provide a valuation of the action
Involved actors	Public and private stakeholders
Expected timeline for action	☐ Short term (1–4 years) ☐ Medium term (5–10 years) ☐ Long term (> 10 years)
Good practices	Reference to actions already implemented
Critical issues	Possible obstacles to the implementation of the action

Table 9.2 List of the "metadata" information included in the technical sheets to search for the action through filters

Metadata	Content
Scope of the action	☐ Mitigation ☐ Adaptation
Type of proposed action	☐ Soft ☐ Green ☐ Gray
Sector of action	☐ Agriculture/forest/land use ☐ Aquaculture/fishery ☐ Biodiversity/ecosystem conservation ☐ Coastal management ☐ Energy ☐ Industry ☐ Public health ☐ Tourism and leisure ☐ Transport and infrastructure ☐ Urban settlement ☐ Waste management ☐ Water resource management ☐ Other
Climate impacts	☐ Change or loss of biodiversity ☐ Coastal erosion ☐ Drought ☐ Fires/wildfires ☐ Floods ☐ Extreme rainfall ☐ Extreme temperatures ☐ Intense wind ☐ Salinization and acidification of water ☐ Other
Implementation scale	☐ Municipality ☐ Region/country ☐ Province ☐ Association of municipalities ☐ Other
Date of the last update	Date of the last revision of the technical sheet
Source	Link to the original source of information

9.3 Data Collection

The catalogue has been realized through comprehensive research starting with the capitalization of the existing adaptation actions already implemented in all the study areas, starting from the analyzed and summarized according to the technical sheet schema.

The second part of the data collection consisted of an additional literature review of existing adaptation actions already implemented outside the pilot areas, always collected according to the technical sheet scheme. These other actions were identified from (i) international, national and local web platforms (e.g., Climate Adapt https://climate-adapt.eea.europa.eu/), (ii) European or international projects (e.g., Interreg IVB project "Future Cities—urban networks to face climate change" http://www.future-cities.eu/), (iii) international, national and local organizations (e.g., Food and Agriculture Organization (FAO) of the United Nations), (iv) webinars from reliable organizations (e.g., 27. Euro-Mediterranean Center on Climate Change (CMCC)), (v) other sources around the world, especially in European coastal areas (e.g., U.S. Department of Agriculture (USDA)).

The data collection was originally conducted in English. Subsequently, the technical sheets were translated into Italian and Croatian through a translation service and with the support of the project partners. This was done in order to make the actions available in the original languages of the pilot areas and the Adriatic coast in general, while maintaining the consistency of the translations. In addition, the English version is also useful to make the actions available to public authorities and other stakeholders from foreign countries, thus extending the results of the work beyond the Adriatic borders.

This work was completed between May 2020 and April 2021 with the help of all the partners involved in the RESPONSe project and coordinated by the Polytechnic University of Marche.

9.4 The Climate Menu

The Climate Menu for Adriatic Regions (http://www.climatemenu.eu/en/) is a free online repository of adaptation actions that can support local policymaking to address the challenges of climate change (Fig. 9.1). The Climate Menu gathers actions that can be implemented at different scales in the context of the Adriatic regions and beyond. It is available in three languages, Croatian, English, and Italian, to increase the outreach.

The main functions of the Climate Menu for the Adriatic Region are as follows. First, it allows online consultation and/or download of technical sheets on climate change adaptation actions. Second, the user can perform a basic search for broad topics, or an advanced search based on metadata such as scope of action, scale of implementation, climate impact and sector of interest. Third, registered users can add new fact sheets. Once logged in, it is possible to fill out a form and add a new climate action to the repository.

The choice of a web platform instead of a printed collection of technical sheets is to make it more accessible to users and easier to update, not only by the web platform administrators, but also by registered users. It also has the advantage of reducing unnecessary paper consumption and waste.

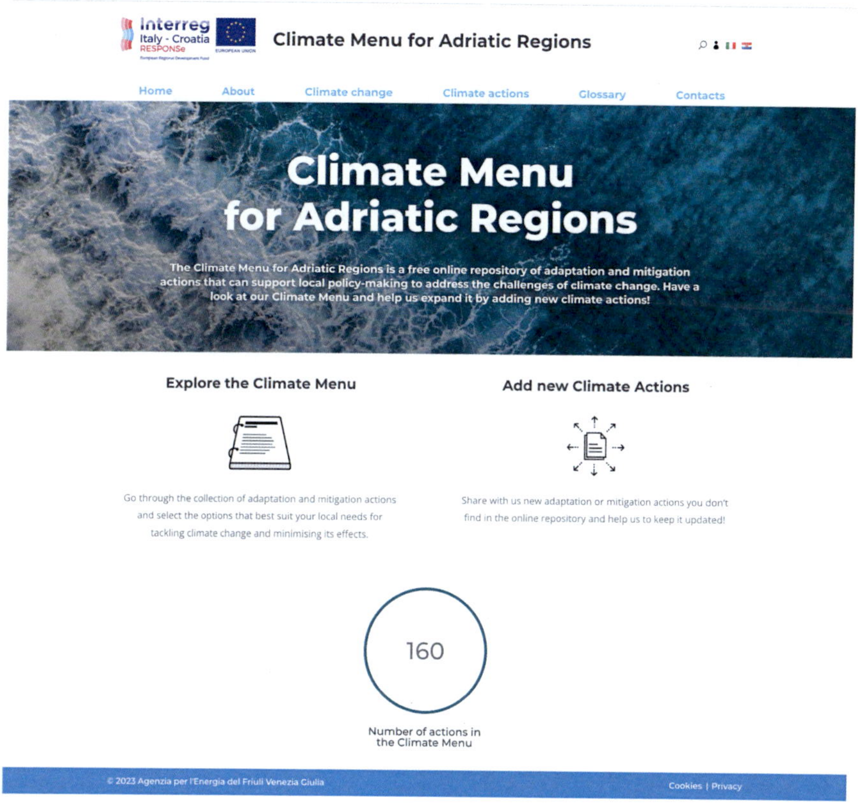

Fig. 9.1 Home page of the Climate Menu for Adriatic Regions web platform. *Source* http://www.climatemenu.eu/en/

9.4.1 An Overview of the Online Database

The tabbed interface of the online database allows navigation to 6 different pages.

The *home* page is the landing page of the web platform (Fig. 9.1). It contains a brief description and the direct link to the main functions of the web platform (exploring the repository and adding new technical sheets).

The *about* page provides users with information about the origins and goals of the Climate Menu, as well as a link to the RESPONSe funding project (Fig. 9.2).

The Climate change page is dedicated to defining climate change, adaptation, and mitigation, as these are the key concepts upon which the Climate Menu is built (Fig. 9.3).

On the Climate actions page, users can search for climate actions through a basic (Fig. 9.4) or advanced (Fig. 9.5) search, selecting specific options according to their needs and interests.

Fig. 9.2 About page of the Climate Menu for Adriatic Regions web platform. *Source* https://www.climatemenu.eu/en/about/

The Glossary page, with more than 200 entries, is designed to help users find definitions of terms related to climate change (Fig. 9.6).

Finally, in the Contacts page, users can find a reference e-mail address to get more information or to send any suggestions to the admins (Fig. 9.7). It is also possible to interact with the admins by filling in a form with comments or requests and sending it directly from this page. This page also displays the logos of the RESPONSe project partners and linked websites.

The cooperation between partners has been essential for opinions' exchange and translating contents and functions' commands and have paved the way to a long-term collaboration, in order to keep the web platform updated in the next future.

9.4.2 *The Adaptation Actions Collected*

A total of 160 technical sheets containing individual adaptation actions were collected. Each technical sheet is linked to at least 2 case studies where the action has already been implemented.

Fig. 9.3 Climate change page of the Climate Menu for Adriatic Regions web platform. *Source* https://www.climatemenu.eu/en/climate-change/

The comprehensive list of technical sheets is available on the online web platform in English, Italian, and Croatian languages. A summary of the database is presented in the following paragraphs.

All the actions collected are mainly intended to adapt to the impacts of climate change, but 28 of them have the synergic effect of also being suitable for the mitigation of greenhouse gas emissions (Fig. 9.8).

The majority of the proposed measures are (i) "green", such as a strategically planned network of natural and semi-natural areas within and around settlements, where a wide range of ecosystem services are provided (e.g., lawns and permeable surfaces, rainwater harvesting systems), followed by (ii) "gray", such as structural and/or engineering solutions (e.g., dams, sea walls, water drainage systems), and

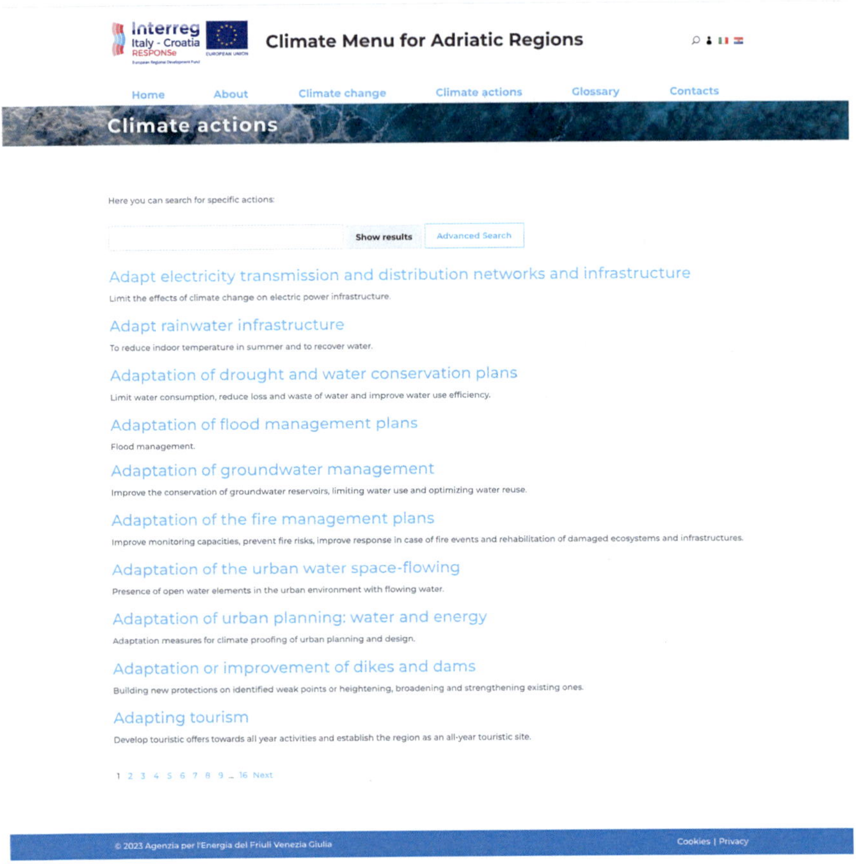

Fig. 9.4 Climate actions basic search page of the Climate Menu for Adriatic Regions web platform. *Source* https://www.climatemenu.eu/en/climate-actions/

(iii) "soft", such as administrative, political, legal, technical, planning, awareness, data collection, monitoring, surveillance, and scientific research solutions (Fig. 9.9). Part of the actions can be attributed to two or even all types of solutions, especially green-soft and gray-soft.

In terms of the sectors targeted by these actions, biodiversity and ecosystem conservation take the lead in representation, closely followed by agriculture, forest management, land use, public health, urban settlements, and water resources management (Fig. 9.10). Other sectors addressed, but not listed in the original structure of the fiches, are for example sport, governance, finance, and activities generally aimed at Disaster Risk Reduction.

Regarding the climate impacts targeted for adaptation, the majority of the compiled actions are geared toward reducing the effects of floods, extreme temperatures, changes or losses in biodiversity, intense rainfall, and drought (Fig. 9.11). Additionally, although not originally listed in the technical sheets, other addressed

Fig. 9.5 Climate actions advanced search within the Climate Menu for Adriatic Regions web platform. *Source* https://www.climatemenu.eu/en/climate-actions/

Fig. 9.6 Glossary page of the Climate Menu for Adriatic Regions web platform. *Source* https://www.climatemenu.eu/en/glossary/

impacts encompass concerns such as sea level rise, snow-related issues, and the management of invasive species.

The envisioned structure for the technical sheets also facilitates the emphasis on the ideal territorial dimensions for implementing these actions. Predominantly, the solutions are tailored for implementation at the municipal scale, with the subsequent focus on regional or county levels, provinces, and finally, at the association of municipalities scale (Fig. 9.12). Furthermore, certain actions demonstrate adaptability across two or more scales, potentially extending to higher or lower levels, contingent upon the intended objectives and beneficiaries.

Finally, the actions can be sorted based on the anticipated timespan for their implementation. The vast majority of these actions are projected to yield results within 1–4 years, closely followed by a nearly equal distribution between those expected to take effect in 5–10 years and those with a horizon of more than 10 years (Fig. 9.13). Similarly, certain actions demonstrate adaptability across multiple timespans, contingent upon their intended objectives and beneficiaries.

9.5 Proposed Adaptation Actions per Each Adriatic Area

In this section, the results of the analysis of the regional strategies, the data collection of the adaptation measures and the assessment of the climate trends are combined in order to define which measures are better suited to the specific needs of each Italian and Croatian region.

Fig. 9.7 Contacts page of the Climate Menu for Adriatic Regions web platform. *Source* https://www.climatemenu.eu/en/contacts/

For this purpose, a filtering process was applied to each study area. The main criteria used were (i) the sectors most involved in the analyzed regional strategies and (ii) the most expected impacts of climate change according to the climate assessment. These criteria served as filters to identify the most appropriate actions. Finally, for each sector, actions were identified that could potentially mitigate all identified impacts.

The following paragraphs present the proposed adaptation actions by region/county. Each action is identified by its corresponding technical sheet number. Notably, the actions highlighted represent those that, if implemented, have the potential to reduce impacts across all three highly engaged sectors. These highlighted actions are identified as key candidates that can address multiple climate impacts from regional to municipal levels of governance.

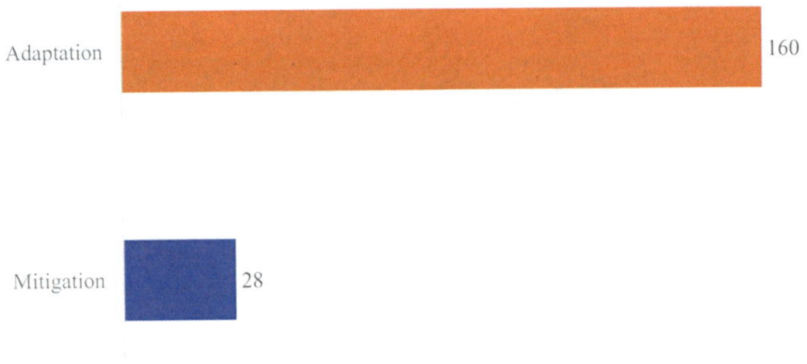

Fig. 9.8 Classification of the collected actions by scope

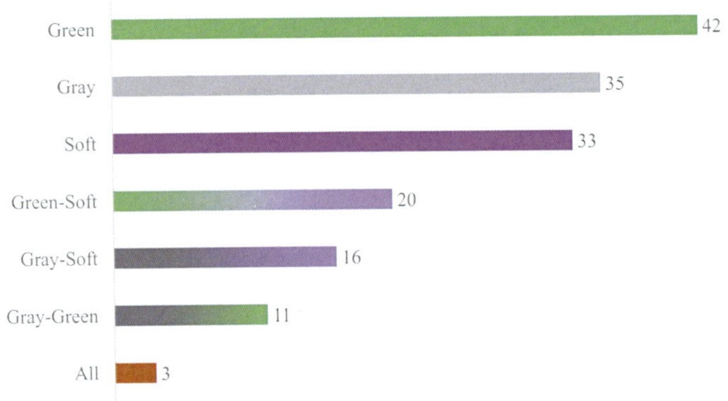

Fig. 9.9 Classification of the collected actions by type: (i) green, (ii) gray, (iii) soft, (iv) green and soft, (v) gray and soft, (vi) gray and green, (vii) all types

In the Friuli-Venezia Giulia Region (Table 9.3), the sectors most affected by the planned regional strategies are biodiversity and ecosystem conservation, energy, and transport and infrastructure. In addition, according to climate trends, the main impacts expected are drought, extreme rainfall and extreme temperatures. As a result, the most appropriate adaptation actions that can address all the above impacts and sectors within the Friuli-Venezia Giulia Region are 2—Capacity building, 83—Information and alerting of the population, 94—Organizing climate workshops, and 102—Promotion of sustainable climate meetings.

In the Primorje-Gorski Kotar County (Table 9.4), the sectors most implicated in the described regional strategies are industry, and transport and infrastructure. Moreover, according to climate trends, the main impacts expected are extreme rainfall, extreme temperatures, and intense wind. As a result, the most appropriate adaptation

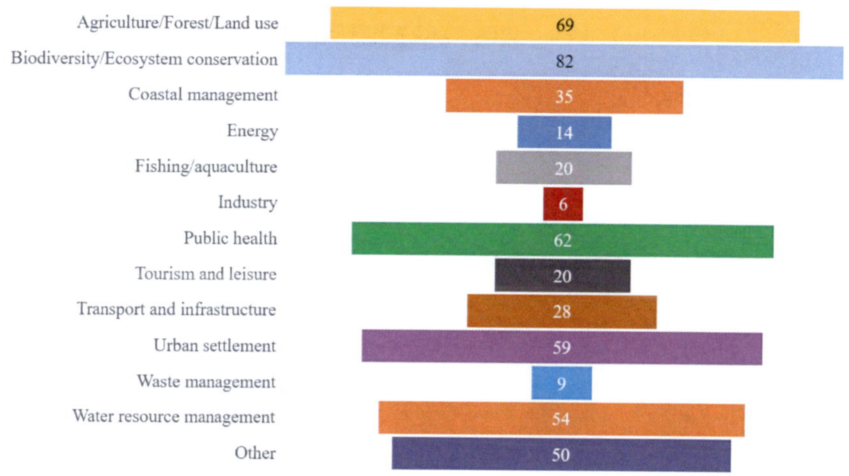

Fig. 9.10 Classification of the collected actions by sector addressed

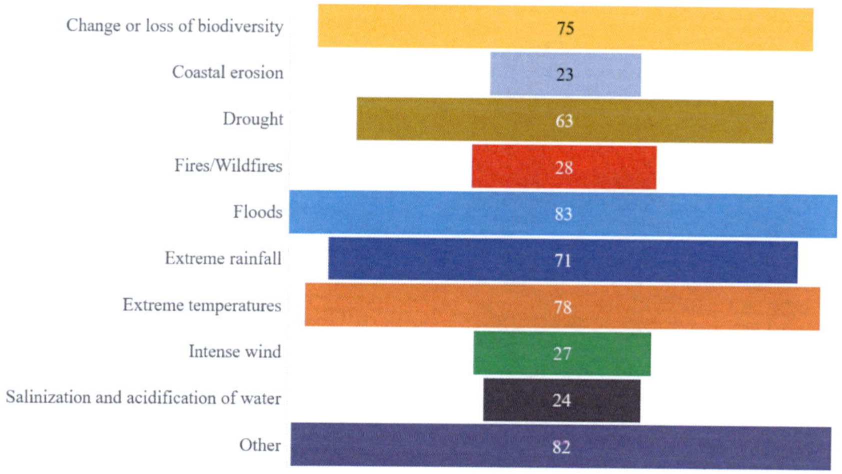

Fig. 9.11 Classification of the collected actions by climate impact

actions that can address all the mentioned impacts and sectors within the Primorje-Gorski Kotar County are 2—Capacity building, 83—Information and alerting of the population, 94—Organizing climate workshops, and 102—Promotion of sustainable climate meetings.

In the Marche Region (Table 9.5), the sectors most interested by the planned regional strategies are agriculture/forest/land use, biodiversity and ecosystem conservation, and urban settlement. In addition, according to climate trends, the main impacts expected are drought, extreme rainfall, extreme temperatures, and intense

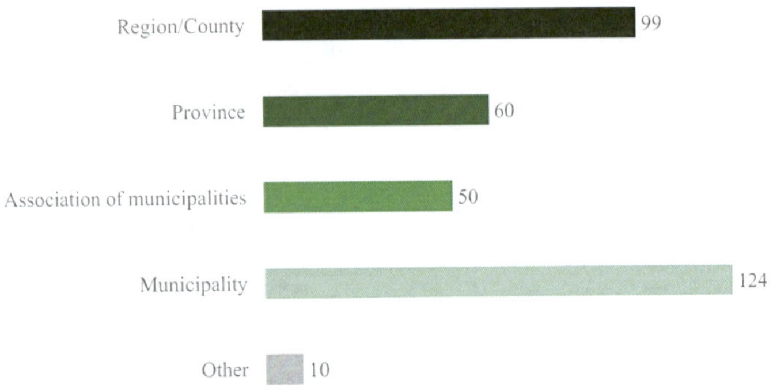

Fig. 9.12 Classification of the collected actions by scale of implementation

Fig. 9.13 Classification of the collected actions by expected timespan

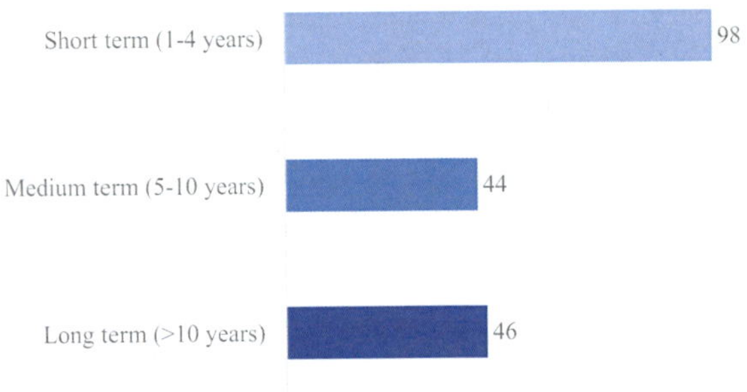

winds. As a result, the most appropriate adaptation actions that can address all of the above impacts and sectors in the Marche Region are 2—Capacity building, 83—Information and alerting the population, 94—Organizing climate workshops, 102—Promotion of sustainable climate meetings, 125—Creating and expanding reserves, 126—Extending the boundaries of reserves, 128—Favoring or restoring native species, 129—Guiding species composition in early stages of stand development, 131—Landscape-scale planning and partnerships, and 138—Preparing for more frequent and severe disturbances.

In the Šibenik-Knin County (Table 9.6), the sectors most affected by the planned regional strategies are fishery/aquaculture, tourism and recreation, and water management. In addition, according to climate trends, the main impacts expected are extreme rainfall and extreme temperatures. As a result, the most appropriate adaptation measures that can address all of the above impacts and sectors in Šibenik-Knin County are 2—Capacity building, 83—Information and alerting the population, 94—Organizing climate workshops, 102—Promotion of sustainable climate meetings,

Table 9.3 Friuli-Venezia Giulia proposed adaptation actions (number) according to the most expected climate change impacts and the sectors most involved in the regional strategies

Most expected impacts	Most engaged sectors	Action number
• Drought • Extreme rainfall • Extreme temperatures	Biodiversity/ ecosystem conservation	**2**-**83**-93-**94**-98-**102**-103–105-106–108-121–123-124–125-126–127-128–129-130–131-132–134-135–136-137–138-158–160
	Energy	**2-83-94-102**
	Transport and infrastructure	**2**-25–69-**83**-90-**94**-**102**-109–125-131–160

Actions that have the potential to reduce impacts in all sectors are highlighted

Table 9.4 Primorje-Gorski Kotar proposed adaptation actions (number) according to the most expected climate change impacts and the sectors most involved in the regional strategies

Most expected impacts	Most engaged sectors	Action number
• Extreme rainfall • Extreme temperatures • Intense wind	Industry	**2-83-94-102**
	Transport and infrastructure	**2**-8–25-**83-94-102**-109–119-125–131

Actions that have the potential to reduce impacts in all sectors are highlighted

125—Creating and expanding reserves, 131—Landscape-scale planning and partnerships, 137—Moving species, and 138—Preparing for more frequent and severe disturbances.

In the Apulia Region (Table 9.7), the sectors most implicated in the planned regional strategies are agriculture/forest/land use, transport and infrastructure, and water resource management. In addition, according to climate trends, the main impacts expected are drought, extreme rainfall and extreme temperatures. As a result, the most appropriate adaptation measures that can address all of the above impacts and sectors in Apulia Region are 2—Capacity building, 83—Information and alerting the population, 94—Organizing climate workshops, 102—Promotion of sustainable climate meetings, 125—Creating and expanding reserves, and 131—Landscape-scale planning and partnerships.

In the Dubrovnik-Neretva County (Table 9.8), the sectors most interested by the planned regional strategies are coastal management, energy, and tourism and leisure. In addition, according to climate trends, the main impacts expected are extreme rainfall and extreme temperatures. As a result, the most appropriate adaptation measures that can address all of the above impacts and sectors within the Dubrovnik-Neretva County are 2—Capacity Building, 83—Information and Alerting of the Population, 94—Organizing Climate Workshops, and 102—Promotion of Sustainable Climate Meetings.

Notably, despite their unique characteristics, all the regions share some commonalities in the actions deemed most effective for fostering adaptation to climate change. Actions such as 2—Capacity building, 83—Information and alerting of the population, 94—Organizing climate workshops, and 102—Promotion of sustainable climate meetings emerge as common threads across the study areas. These actions seem to hold significance as universally beneficial strategies for adapting to climate change across diverse territories. Specifically, they are all "soft" adaptation actions, except for 94—Organizing climate workshops, which is also indicated for mitigation. As "soft" measures, they can be applied to multiple sectors and address multiple climate impacts at different levels of governance.

The action 2, titled "Capacity Building", aims to invest in the effectiveness and future sustainability of society by employing various methods. These methods encompass diverse forms of capacity building processes, spanning education through schools, universities, and other educational services, training via courses, seminars, and e-learning, as well as networking opportunities such as conferences, workshops, and communities of practice. Additionally, specific coaching, technical assistance

Table 9.5 Marche proposed adaptation actions (number) according to the most expected climate change impacts and the sectors most involved in the regional strategies

Most expected impacts	Most engaged sectors	Action number
• Drought • Extreme rainfall • Extreme temperatures • Intense wind	Agriculture/forest/land use	**2-83-94-102**-124-**125-126**-127-**128-129**-130-**131**-132–134-135–136-137-**138**-158
	Biodiversity/ecosystem conservation	**2-83-94-102**-123–124-**125-126**-127-**128-129**-130-**131**-132–134-135–136-137-**138**-158
	Urban settlement	**2-83-94-102-125-126-128-129-131-138**

Actions that have the potential to reduce impacts in all sectors are highlighted

Table 9.6 Šibenik-Knin proposed adaptation actions (number) according to the most expected climate change impacts and the sectors most involved in the regional strategies

Most expected impacts	Most engaged sectors	Action number
• Extreme rainfall • Extreme temperatures	Fishery/aquaculture	**2-83-94-102-125-131**-132-**137-138**
	Tourism and leisure	**2-83**-90-**94-102-125**-126–128-**131**-134-**137-138**
	Water resource management	**2**-3-**83**-85-**94**-101-**102**-103–104-105–106-107–108-110-**125**-130-**131**-132-**137-138**

Actions that have the potential to reduce impacts in all sectors are highlighted

Table 9.7 Apulia proposed adaptation actions (number) according to the most expected climate change impacts and the sectors most involved in the regional strategies

Most expected impacts	Most engaged sectors	Action number
• Extreme rainfall • Extreme temperatures • Drought • Change or loss of biodiversity	Agriculture/forest/land use	**2**-3-**83**-85-**94**-**102**-108–121-124-**125**-126–127-128–129-130-**131**-132–134-135–136-137–138-158
	Transport and infrastructure	**2**-25–69-**83**-90-**94**-**102**-109-**125**-**131**-160
	Water resource management	**2**-3-**83**-85-**94**-101-**102**-103–105-106–108-110-**125**-130-**131**-132–137-138

Actions that have the potential to reduce impacts in all sectors are highlighted

Table 9.8 Dubrovnik-Neretva proposed adaptation actions (number) according to the most expected climate change impacts and the sectors most involved in the regional strategies

Most expected impacts	Most engaged sectors	Action number
• Extreme rainfall • Extreme temperatures	Coastal management	**2**-3-**83**-**94**-**102**-125–131-160
	Energy	**2**-**83**-**94**-**102**-104
	Tourism and leisure	**2**-**83**-90-**94**-**102**-125–126-128–131-134–137-138

Actions that have the potential to reduce impacts in all sectors are highlighted

like expert missions, twinning, and attention to groups at risk are integral components. The envisioned outcome of these efforts is to bolster sustainability and institutionalize knowledge regarding climate change adaptation. This objective is quantified through indicators like the number of informed individuals engaged in these initiatives. The action primarily involves communication professionals and the wider population, aiming for short-term results within a 1–4 year timeframe.

However, a critical challenge lies in serving diverse target groups with distinct knowledge needs. Even when information is available, adaptation and relevance to suit these varying audiences remain pivotal for successful implementation and impact.

Several good practices around the world, showcasing successful implementation of similar strategies, include London (UK) where social housing landscapes were climate-proofed by Groundwork London and Hammersmith and Fulham Council, Tatabánya (Hungary) addressing urban heat waves and forest fires through alert measures, and Norway's utilization of insurance loss data by local authorities. Moreover, Marche Region and Šibenik-Knin County, both integrate such action on local planning strategies for climate change adaptation.

The action 83, titled "Information and alerting of the population" has a core objective of ensuring the safety of individuals, infrastructures, buildings, and economic activities during emergencies. This involves a continuous flow of information, covering expected or ongoing phenomena, specific self-defense measures, evolving situations, rescue operations, contact details, and post-event guidance such as evacuation procedures. Various communication channels are employed for disseminating this critical information. These include traditional methods like speakers mounted on vehicles, sirens, as well as modern platforms such as dedicated websites, social networks, and text messages. The anticipated outcome is an informed population equipped with knowledge about appropriate actions to take during and after an emergency. The success of this action is gauged through metrics such as the number of individuals reached by these alert systems. The involved actors primarily consist of civil protection agencies, local government bodies, and citizens themselves. The action plan targets short-term implementation within a 1–4 year period.

However, a significant challenge lies in ensuring universal access to these channels. The methods employed for communication may not be suitable for all citizens, potentially leading to information gaps where certain demographics might not receive critical alerts and guidance during emergencies. This issue highlights the importance of diverse communication strategies to reach all segments of the population effectively.

Several examples of successful practices in this domain include Vicenza in the Italian Veneto Region, the United States' Federal Emergency Management Agency (FEMA) with its Integrated Public Alert Warning System (IPAWS), Bolzano in the Italian Trentino-Alto Adige Region, and initiatives in France, emphasizing the importance of alerting and informing populations during emergencies.

The action 94, titled "Organizing Climate Workshops" aims to engage individuals about climate change through collaborative events and exchanges. These workshops or twinning programs serve as platforms for discussing climate-related topics, devising measures and plans for development, exchanging experiences, and seeking advice across international borders. The expected outcome is the dissemination of information to various groups, including administrations, students, and investors, to influence awareness and attitudes toward climate change. The primary goal is to facilitate problem-solving through direct, practical exchanges and consultations among local/regional authorities and experts from diverse countries, regions, or localities. The success of these efforts is quantified by the number of participants engaged in these workshops. Key actors involved in this action encompass public administrations, professional experts, citizens, universities, and associations. The action plan sets a short-term timeline for implementation, typically within 1–4 years.

However, two critical challenges impede the effectiveness of these workshops: funding constraints and communication issues. Securing adequate funding for organizing and sustaining these workshops poses a challenge, potentially limiting their reach and impact. Additionally, communication barriers might hinder the dissemination of information to the wider audience, emphasizing the need for robust and inclusive communication strategies to ensure broader participation and knowledge dissemination.

Several best practices of such initiatives include examples in the United States, Venice in the Italian Veneto Region, and Zagreb in Croatia. Also, among the study areas, this action has been tested in Pordenone (Friuli-Venezia Giulia Region), Ancona (Marche Region), Bari (Apulia Region). These events and collaborations have showcased innovative strategies to respond to climate change, often focusing on agriculture, water management, coastal protection, and urban planning.

The initiative 102, titled "Promotion of sustainable climate meetings", centers on providing comprehensive information on climate change through organized conventions. These meetings focus on various climate-related topics, such as storms, droughts, temperature changes, precipitation patterns, wind dynamics, and sea level changes. The primary goal is to facilitate informed discussions that enable proactive action, with an emphasis on investing in technical and potentially expensive solutions, such as higher levees or expanded sewer systems, to address climate challenges. The expected outcome is the empowerment of participants with the ability to take proactive measures against climate issues, particularly by investing in infrastructure and technical solutions. The success of these efforts is measured by the number of participants engaged in these climate meetings. Crucial actors involved in these endeavors include local authorities, climate experts, technicians, and citizens. The action plan is set within a short-term framework, typically spanning 1–4 years.

However, two critical challenges hinder the efficacy of these meetings: space availability and the cost associated with organizing such events, coupled with the challenge of involving a diverse range of people. The limitations in available space for hosting these gatherings may restrict participation, while the financial burden could impede the inclusivity and scale of these meetings, potentially limiting the breadth of engagement and collaboration among stakeholders. Addressing these challenges requires creative solutions to ensure accessibility and active involvement from a diverse set of participants.

Successful practices in this domain have been observed in various regions, including the Emscher Region in Germany and multiple regions within Italy such as the study areas of Friuli-Venezia Giulia, Marche, and Apulia. These regions have showcased their approaches to climate change adaptation and mitigation through comprehensive reports, conventions, and regional strategies.

In addition, as a result of this analysis, two further common actions for adaptation to climate change in the central Adriatic macro-region and the Apulia region were proposed, namely 125—Creating and expanding reserves and 131—Landscape-scale planning and partnerships. These adaptation actions are both "green" and "soft" and cover all climate change impacts and almost all sectors, with the exception of energy, industry and waste management.

The action 125, titled "Creating and expanding reserves" aims to create and enlarge natural reserves or networks of reserves to connect habitats and safeguard critical ecosystems. These reserves are selected based on natural areas with minimal human interference, allowing for natural disturbance processes like fire management while ensuring limited or no harvesting activities. The initiative involves positioning reserves adjacent to each other to form networks consisting of large reserves, numerous small reserves across a latitudinal range, or a mix of both, to maintain connectivity across diverse landscapes. For this reason, it can be applied only at the regional level of governance or more. The anticipated outcome includes increased flexibility in land management practices adjacent to these reserves, facilitating species movement, and safeguarding core areas from disturbances that could jeopardize their ecological integrity. The success of these efforts is gauged by the number of plant and animal species sheltered within these protected areas. Key stakeholders involved in this action encompass government bodies, international/national/regional/local donors, environmental experts, and local communities. The action plan spans short-term (1–4 years), medium-term (5–10 years), and long-term (> 10 years) timelines.

However, a critical challenge lies in the vulnerability of plant and animal species within these reserves. Despite their protected status, several factors such as habitat degradation, climate change, and human activities can still threaten the well-being of these species, requiring ongoing monitoring and adaptive management strategies to ensure their long-term survival within these reserves.

Successful examples for establishing and expanding reserves have been observed in various countries like the Dominican Republic, Madagascar, Ecuador, Indonesia,

New Zealand, Guinea-Bissau, and Canada. Additionally, several regions within Italy, including Friuli-Venezia Giulia, Veneto, Marche, Emilia-Romagna, and Apulia, as well as Croatia, have implemented strategies to protect natural reserves and parks.

The action 131, titled "Landscape-scale planning and partnerships" focuses on employing landscape-scale planning and collaborative partnerships to mitigate fragmentation and enhance connectivity across various ecosystems. This approach targets terrestrial, aquatic, or urban/peri-urban landscapes on different scales, including the field, farm, landscape, and regional levels. Enhancing connectivity involves various strategies such as creating linear habitat strips, preserving field margins, conserving wooded areas near water bodies, and establishing hedgerows and windbreaks. These corridor systems not only foster landscape connectivity but also serve additional conservation objectives, like safeguarding Crop Wild Relatives (CWR) which are wild plants closely related to cultivated crops. The expected outcomes encompass a wide array of benefits, including carbon sequestration, species conservation, mobility of food/habitat/species, gene flow, pollinator support, pest control, soil conservation, water regulation, air quality improvement, positive business reputation, community engagement platforms, employee benefits, operational licenses, recreational value, aesthetic enhancements, ecotourism opportunities, collaboration between public administration and academic/research groups, and improvements in human health and well-being. The measurement of success is tied to the number of endemic and non-endemic plant and animal species that thrive within these connected landscapes. Key stakeholders involved in this initiative span governments, public administrations, scientists, and citizens. The action plan operates within short-term (1–4 years) timelines.

However, a critical challenge lies in fostering cooperation among diverse stakeholders like governments, public administration bodies, scientists, and citizens. Achieving consensus and effective collaboration among these varied entities is crucial for successful landscape-scale planning and execution, as differing priorities and approaches can hinder the seamless implementation of connected landscape initiatives. Addressing these issues requires effective communication, shared goals, and coordinated efforts among all involved parties.

Best practices for landscape-scale planning and connectivity have been observed globally, including practices in California (USA), Delhi (India), Nagoya (Japan), and the UK. These examples highlight the effectiveness of collaborative efforts in preserving and enhancing landscape connectivity for biodiversity and ecological sustainability.

Moreover, the action 138, titled "Preparing for more frequent and severe disturbances" is proposed for both the study areas of central Adriatic, namely Marche Region and Šibenik-Knin County. Such adaptation action is both "green" and "soft", can be implemented at all levels of governance, cover all climate change impacts and several sectors, including agriculture/forest/land use, fishery/aquaculture, biodiversity/ecosystem conservation, public health, tourism and leisure, urban settlement, and water resource management.

The initiative is aimed at proactively responding to increasingly frequent and severe disruptions caused by climate change. The action involves documenting clear plans in advance to enable faster, more thoughtful, and better-coordinated responses to these disturbances. This proactive approach includes identifying areas where certain forest types might struggle to regenerate after severe disturbances and devising response strategies, such as establishing better-adapted communities in these vulnerable locations. Considerations for adaptation actions involve understanding the impact of stand-replacing disturbances on different forest types. For instance, aspen and paper birch forest types can benefit from such disturbances, while balsam fir and spruce forest types are highly susceptible to damage from fire and wind due to their flammable needles and shallow roots. Jack pine forests, on the other hand, require significant site disturbance to regenerate and are typically favored by stand-replacing disturbances like recurring fires. The expected outcomes revolve around planning and studying the primary drivers of disturbance events, including drought, dry soil and peatlands, wildfires, invasive species, and severe weather events. Success in this area is measured by monitoring the frequency and intensity of these disturbance events over time. Key actors involved in implementing these plans include scientists, natural planners, natural risk experts, and government and public administration bodies. The action plan operates within a long-term (> 10 years) timeline.

However, critical challenges arise from the impacts of climate change itself, such as drought, dry soil and peatlands, wildfires, invasive species, and severe weather events. Addressing these issues requires ongoing assessment, adaptive planning, and responsive actions to cope with and mitigate the impacts of these disturbances on ecosystems and communities.

Several regions and countries, such as Alaska, the USA, Canada, and the Mediterranean area, have implemented good practices in addressing disturbance ecology and adaptation strategies. These examples highlight the importance of understanding disturbance impacts and developing proactive response plans to mitigate the adverse effects of climate change-induced disruptions.

Conversely, in terms of regional specificities, the most appropriate adaptation actions that can address climate change in the Marche Region are 126—Extending the boundaries of reserves, 128—Favoring or restoring native species, and 129—Guiding species composition in the early stages of stand development. These adaptation actions are mostly "green", apart from action number 126, which is both "green" and "soft". They cover all climate change impacts, except coastal management, and many sectors, including agriculture/forest/land use, biodiversity/ecosystem conservation, public health, tourism and leisure, and urban settlement.

The action 126, titled "Extending the boundaries of reserves", is designed to safeguard ecosystems by extending the boundaries of existing reserves and protected ecosystems. These protected areas, typically defined as natural areas with minimal human interference and open to natural disturbance processes like fire management, serve as crucial habitats for various species. The expansion of reserve boundaries serves multiple purposes. It not only buffers and replicates diversity within the core of the reserve but also broadens the overall variation in species within the expanded area. This proactive step aims to bolster species diversity and protect ecosystems from external threats. The expected outcome revolves around an increased richness and diversity of species within the expanded reserve boundaries. Success in this initiative is measured by the number of species preserved within these extended protected areas. It can be implemented at all levels of governance. Key stakeholders involved in implementing this action include governments, natural reserve authorities, reserve visitors, and national park services. The action plan spans short-term (1–4 years), medium-term (5–10 years), and long-term (> 10 years) timelines.

However, critical challenges persist due to factors like high population density, urban demand for charcoal and fuelwood, and the general population's demand for forest resources. Balancing conservation efforts with the growing demands on natural resources necessitates comprehensive strategies that consider both environmental preservation and human needs. Addressing these challenges requires sustainable resource management approaches, community involvement, and policies that balance conservation goals with socio-economic realities.

Successful practices in expanding reserve boundaries have been observed in various regions, such as the West Mengo Region in Uganda, the Portland metropolitan Region in Oregon (USA), and the Philippines. These examples showcase the effectiveness of enlarging protected areas to safeguard diverse ecosystems and preserve species diversity.

The action 128, titled "Favoring or restoring native species", aims to promote or reintroduce native species anticipated to be better suited to future climate conditions. This approach involves managing ecosystems to facilitate the dominance of native species that are expected to thrive or adapt well in the face of changing climatic conditions. The strategy entails various methods such as favoring native species in communities or forest types that are projected to benefit under future climate scenarios. This management may lead to a gradual shift in forest composition, creating opportunities for future-adapted species to occupy niches left by declining species. For instance, planting a variety of native species to increase overall species richness or emphasizing more drought- and heat-tolerant species in areas expected to become warmer and drier. For this reason, such action should be implemented at the regional level of governance. The expected outcomes involve identifying species capable of withstanding and thriving under future climate conditions, including harsh urban environments. Success is evaluated by the preservation and establishment of native species suited to changing climatic conditions. Key stakeholders involved in this action encompass animal or plant experts, natural managers, communities, and scientists. The action plan operates within a medium-term (5–10 years) and long-term (> 10 years) timeline.

However, critical challenges persist due to variations in species tolerance to future climatic conditions. Different species have diverse capacities to adapt, and ensuring the success of native species promotion relies on understanding these nuances and effectively managing ecosystems to facilitate their growth and resilience in changing environments. Addressing these challenges requires ongoing research, adaptive management, and informed decision-making to support the preservation and restoration of resilient native species.

Effective practices in favoring or restoring native species have been observed in various regions, such as the Caribbean and various parts of the USA (Midwest and

Northwest). These instances highlight the importance of promoting native species resilient to future climatic conditions for ecological adaptation and sustainability.

The action 129, titled "Guiding species composition in early stages of stand development", aims to guide forests toward new and better-adapted compositions, particularly in response to natural disturbances and changing environmental conditions. This approach leverages natural disturbances to initiate seedling development and genetic mixing, facilitating adaptation in forests. Silvicultural practices are employed to replicate natural disturbances and promote regeneration in the absence of such natural events. Various techniques are applied in this approach, including preventing and removing undesired species, controlling aggressive native or invasive non-native species, and conducting timber stand improvement to foster the growth of desirable species. These actions aim to promote desired species and minimize competition from poorly adapted or invasive species, ultimately leading to a potential transition to a different forest type better suited to changing conditions. For this reason, such action should be implemented at the regional level of governance. The expected outcomes involve the promotion of desired species and the reduction of competition from undesired, poorly adapted, or invasive species. Success is measured by the regeneration of forests and the establishment of desired species. Key stakeholders involved in this action include natural managers, ecosystem experts, government bodies, and local communities. The action plan operates within a medium-term (5–10 years) and long-term (> 10 years) timeline.

However, critical issues arise due to drier conditions and increased stress on forest ecosystems. Addressing these challenges requires adaptive management strategies, continuous monitoring, and research to mitigate the impacts of changing environmental conditions on forest health and development. Implementing proactive measures to guide forest composition amid changing conditions is crucial for enhancing ecosystem resilience and sustainability.

Effective practices in guiding species composition at early stand development stages have been observed in various regions, such as Canada, Catalonia in Spain, and Hokkaido in Japan. These good practices highlight the efficacy of employing silvicultural techniques to guide forest development toward more resilient and better-adapted compositions.

Finally, the action 137, titled "Moving species", is proposed only for Šibenik-Knin County. Such adaptation action is "green", should be implemented at the regional level of governance, cover all climate change impacts, except coastal management, and several sectors, including agriculture/forest/land use, biodiversity/ecosystem conservation, fishery/aquaculture, tourism and recreation, and water resources management.

The initiative focuses on the identification and relocation of species to areas anticipated to offer suitable habitat in the future. This strategy involves maintaining ecosystem functionality or transitioning to better-adapted systems by actively introducing non-native species or specific genetic variations. Given the uncertainty surrounding future climate conditions, success in this approach may hinge on relocating species with a wide range of tolerances across various provenances, encompassing different climatic and environmental conditions. The expected outcome involves establishing populations of species well-suited to specific climatic and site conditions in new habitats. The success of these efforts is measured by the number of species relocated to these targeted areas. Key stakeholders involved in executing this action include natural managers, scientists, and ecologists. The action plan operates within a medium-term (5–10 years) and long-term (> 10 years) timeline.

However, critical challenges stem from the uncertainties inherent in climate change predictions, the limited track record of previous species relocations, and ongoing uncertainties regarding ecosystem responses. The dynamic nature of ecosystems and the unpredictability of climate change impacts pose significant challenges, necessitating continual assessment, adaptive management, and careful monitoring of the relocated species success in their new habitats. Addressing these challenges will require ongoing research, collaboration among experts, and adaptive strategies to effectively manage and mitigate uncertainties in species relocation programs.

Successful practices in relocating species have been observed in various regions, including Australia, the USA, and the Czech Republic. These instances have highlighted the potential of relocating species as an adaptive strategy to ensure their survival and adaptation to changing climatic conditions.

9.6 Summary

The Climate Menu for Adriatic Regions toolkit represents the synthesis of the conceptual framework presented in this book, aimed at promoting adaptation to climate change in the Adriatic region. The Climate Menu is built through an initial capitalization of actions already implemented in the study areas, and a subsequent literature review of existing adaptation actions already implemented outside these areas.

Initially, the focus of the analysis revolved around the regional strategies collected in the previous phases, recognizing their amalgamation into broader European strategies through a hierarchical structure. The aim was to delve into the precise issues and practices already in place in the Adriatic regions. This was followed by a comprehensive integration of additional actions, drawing on policies and initiatives implemented within Europe and extending beyond its borders. This approach aimed to build a comprehensive understanding of effective climate change strategies relevant to the Adriatic. Finally, in order to prioritize adaptation efforts based on the tangible impacts of climate change, a careful selection of actions capable of addressing current and future climate trends in the Adriatic macro-regions was undertaken. This step aimed at aligning efforts with the current and expected impacts of climate change in the region.

The Climate Menu consists of 160 separate technical sheets that present the individual adaptation actions, and in some cases both adaptation actions, that have been collected. The technical sheets provide information on the objective of the measure, the main expected results, the actors to be involved, the expected outcomes, the timing of implementation, the scope and type of measure, the sectors that can benefit from the implementation of the measure and the climate change impacts that can be reduced, and the scale of implementation. In addition, each fact sheet contains a reference to two or more case studies, so that the end user, i.e., a public authority, has an example of the actual benefits that can be derived from the application of the measure contained in the fact sheet.

Looking more closely at the specifics of the technical sheets, the actions collected are mainly aimed at adapting to climate change, although 28 technical sheets have the synergistic effect of also being suitable for mitigating climate change. The majority of the proposed actions are (i) "green", followed by (ii) "gray" and (iii) "soft" actions. The sectors most addressed by the actions are biodiversity/ecosystem conservation, followed by agriculture/forest/land use, public health, urban settlements and water resources management. In terms of the climate impacts to which adaptation is targeted, the majority of the actions collected are proposed to adapt to the impacts of flooding, extreme temperatures, changes or loss of biodiversity, extreme rainfall and drought.

In the next phase, tailored adaptation measures were proposed for each of the different study regions in Italy and Croatia. For example, in Friuli-Venezia Giulia, adaptation measures such as capacity building, population alert systems, climate workshops and sustainable meetings were considered appropriate to address the impacts of drought, extreme precipitation and high temperatures on biodiversity, energy and transport. In Primorje-Gorski Kotar County, the proposed adaptation measures include capacity building, community awareness programs, climate workshops and sustainable meetings. These targeted actions aim to mitigate impacts in the industrial, transport and infrastructure sectors, and to adapt to the expected climate changes in the region, i.e., extreme rainfall, temperatures and strong winds. The Marche Region adaptation actions address agriculture/land use, biodiversity, urban settlement, and expected climate impacts such as droughts, extreme precipitation, temperatures, and intense winds. Actions include capacity building, community awareness raising, climate workshops, sustainability meetings, creation/expansion of protected areas, conservation of native species, landscape planning, and preparation for severe disturbances. The adaptation measures for Šibenik-Knin County are tailored to fisheries/aquaculture, tourism, water management and the expected impacts of extreme precipitation and temperatures. The proposed actions revolve around capacity building, public awareness, climate workshops, sustainable meetings, creation/expansion of protected areas, landscape planning, species translocation, and preparation for severe disturbances. The Apulia Region strategies revolve around agriculture/land use, transportation, water management, and expected impacts, including drought, extreme precipitation, and temperature. Proposed adaptation measures include capacity building, community awareness raising, climate workshops, sustainable meetings, creation of protected areas, and landscape planning. In Dubrovnik-Neretva County, adaptation focuses on coastal management, energy, tourism, and the projected impacts of extreme rainfall and temperatures. The proposed measures include capacity building, community awareness raising, climate workshops, sustainable meetings, and measures specific to coastal management, energy systems, and tourism activities.

Among the 160 possible actions that can be implemented in the Adriatic Basin to adapt to climate change, 2—Capacity building, 83—Informing and alerting the population, 94—Organizing climate workshops, and 102—Promoting sustainable climate meetings emerge as common threads across the study areas. Two other common actions for adaptation to climate change in the central Adriatic macro-region and the Apulia Region were proposed, namely 125—Creation and expansion of protected areas and 131—Landscape-scale planning and partnerships. In addition, for the two study areas of the central Adriatic, i.e., the Marche Region and the Šibenik-Knin County, the action 138—Preparing for more frequent and more severe disturbances is proposed. Conversely, in terms of regional specificities, the most appropriate adaptation measures that can address climate change in the Marche Region are 126—Extending the boundaries of protected areas, 128—Promoting or restoring native species, and 129—Guiding species composition in the early stages of stand development. Finally, action 137—Moving species is proposed only for Šibenik-Knin County.

All these actions imply the use of "soft" and "green" strategies and cover a wide range of initiatives aimed at reducing the impacts of climate change across different sectors and levels of governance. Each action reflects a strategic approach tailored to the unique characteristics and vulnerabilities of each region. The diversity of challenges, from financial constraints to communication barriers and uncertainties in climate projections, underscores the complexity of addressing climate change. However, the actions identified provide a roadmap for regions to proactively adapt, engage stakeholders, and promote sustainable strategies to mitigate the adverse effects of climate change. Through initiatives such as expanding protected area boundaries, promoting native species, managing forest composition, creating protected areas, and organizing workshops, these regions are positioning themselves to address the multifaceted challenges posed by climate change. The examples presented from different countries and regions offer valuable insights and best practices, providing a rich repository of successful strategies that can serve as models for these regions' adaptation efforts. In addition, the emphasis on collaboration among stakeholders-governments, local communities, scientists, and environmental experts-underscores the importance of collective action and knowledge sharing in addressing the challenges of climate change.

The comprehensive analysis and proposed actions underscore a holistic approach to climate change adaptation that recognizes the need for multifaceted strategies across sectors and levels of governance. These initiatives, if effectively implemented, could contribute significantly to building resilience and sustainability in the face of evolving climate conditions.

Finally, in order to guide local authorities in the selection of possible actions that can be implemented in their territory to mitigate and adapt to climate change, the 160 actions collected have been gathered in the Climate Menu online platform (http://www.climatemenu.eu/en/). The online platform has a double advantage: (i) it is accessible to all users who need to consult the collected actions or want to add a new action, thus allowing a continuous and constant update; (ii) the platform itself, by preferring the online design to the printed one, is a first step in combating climate change by reducing unnecessary paper consumption and waste. Furthermore, the online nature of the platform allows it to be consulted by citizens, local authorities, organizations and associations interested in defining a strategy to combat climate change.

In summary, the Climate Menu stands as a crucial tool capable of steering public authorities in crafting robust planning policies for both mitigation and adaptation. It empowers decision-making processes by intricately weaving these policies into the fabric of local needs and the unique characteristics of their respective macro-areas.

Moving forward, the next phase of this process should entail engaging local stakeholders, leveraging a bottom-up approach. This consultation process holds immense potential to further enrich and refine this work. By involving stakeholders at the grassroots level, this initiative stands to gain valuable insights and support, ultimately enhancing its effectiveness and relevance within the communities it aims to serve.

Chapter 10
Conclusions: Pathways to Climate Adaptation in the Adriatic Coastal Areas

The culmination of this comprehensive exploration into climate change adaptation in the Adriatic coastal regions underscores the intricate and multifaceted nature of addressing the challenges posed by a changing climate. This book has meticulously examined the current state of adaptation and mitigation strategies, synthesized climate data with these strategies, and presented a robust toolkit for future actions. The following sections present the key insights and propose a pathway for enhanced resilience and sustainability in the studied area.

10.1 State-of-the-Art Adaptation (and Mitigation) Strategies

The multiscale analysis of climate change counteracting mechanisms allowed for the identification of interlinkages between policies at all levels, from European to national and regional. The study collected the information for 38 adaptation and mitigation strategies: 2 European, 5 national (3 Italian and 2 Croatian), and 33 regional (30 Italian and 3 Croatian). In particular, the northern regional strategies analyzed were 14 (13 Italian and 1 Croatian), the central regional strategies were 11 (10 Italian and 1 Croatian), and the southern regional strategies were 8 (7 Italian and 1 Croatian).

In response to the UN summit, the European Commission has implemented two strategies at the European level, one designed to facilitate climate adaptation and the other intended to promote emission mitigation. The adaptation strategy is designed to foster the resilience of Europe and to develop guidelines, while addressing knowledge gaps through the European Climate Adaptation Platform (Climate-ADAPT). The mitigation strategy has the objective of reducing greenhouse gas emissions by 20% by 2020 and 40% by 2030. This allows companies to trade emission allowances. Both strategies address vulnerable sectors, including industry, agriculture, fisheries, transportation, and infrastructure. The strategies employ a combination of "green"

E. Gioia, *Climate Change Adaptation in the Adriatic Coastal Areas*, Disaster Risk Reduction, https://doi.org/10.1007/978-981-97-6655-0_10

and "soft" measures, with the inclusion of "gray" measures for adaptation. Economic and financial instruments are employed in both cases, with additional voluntary and communication tools for adaptation. The implementation of the strategy is ongoing, with European funding, and the evaluation will involve public consultation and assessments by the EU Commission. The strategies' strengths include cooperation and emission reductions, while their weaknesses include slower local progress and carbon leakage.

The national strategies analyzed are in alignment with the EU adaptation and mitigation strategies, with the specific plans for each territory being tailored to the particular needs of each country. The Italian and Croatian adaptation national strategies show analogous characteristics, with the objective of preparing their respective nations in the face of climate change. The mitigation strategies focus on reducing greenhouse gas emissions by promoting the use of renewable energy and energy efficiency. Both countries' strategies encompass a substantial number of vulnerable sectors and adopt a top-down approach. Italy additionally addresses transboundary risks. The aforementioned strategies have a long-term horizon and incorporate a range of measures, including those of a "gray," "green," and "soft" nature. The strategies are funded by the EU and involve extensive monitoring to evaluate activities.

Similarly, the regional strategies align with European and national guidelines, adapting them to the specific territories. Looking at the north, central and south Adriatic regions, both common and distinct patterns emerge in terms of strategy implementation, reliance on climate projections, sectors covered, policy instruments used, stakeholder involvement, and integration into existing policies.

In the north Adriatic, the Region of Friuli-Venezia Giulia emphasizes both mitigation and adaptation, while the county of Primorje-Gorski Kotar focuses only on mitigation, with detailed projections and vulnerability assessments playing a crucial role in strategy development. Interestingly, despite the geographical location, most of the strategies in the Friuli-Venezia Giulia Region do not consider cross-border risks, but rather involve a wide range of stakeholders in their implementation. In the central Adriatic, the Marche Region and the Šibenik-Knin County comprehensively integrate climate scenarios and vulnerabilities and comprehensively address cross-border impacts through cooperation between public authorities, scientific institutions, associations, private bodies, and civil society. However, the Marche strategy mostly addresses both mitigation and adaptation, while Šibenik-Knin only addresses adaptation. In the south Adriatic, particularly in the Apulia Region and Dubrovnik-Neretva County, the strategies both showed a stronger focus on mitigation, with varying degrees of reliance on climate projections and vulnerability assessments, and no consideration of transboundary risks. In these regions, particularly Apulia, stakeholder engagement is practiced, but with less reliance on research and scientific institutions.

All strategies give importance to several vulnerable sectors, but the distribution is different along the northern, central and southern Adriatic. In the north, the Friuli-Venezia Giulia Region focuses on sectors such as agriculture, biodiversity, electricity, flooding, and infrastructure. However, there are gaps in the areas of emergency services and foreign policy. Primorje-Gorski Kotar County prioritizes industry and transport. In the center, Marche County spans all sectors with a focus on biodiversity, land use, liability and insurance, and urban settlement. Šibenik-Knin County's strategy includes agriculture, biodiversity, fisheries, flooding, land use, liability and insurance, tourism, and water management. In the south, Apulia Region emphasizes various actions in all sectors, especially agriculture, human health, transport and water management. Dubrovnik-Neretva County's strategy covers several sectors, namely agriculture, biodiversity, coastal management, electricity, fisheries, human health, industry, land use, tourism, transport, and water resources management.

The time horizons of these strategies vary, with a common emphasis on "long-term" planning in all regions, with a few exceptions. While the Italian regions focus mainly on "gray" or "green" measures, the strategies in Croatia tend to focus more on "soft" measures, showing a diversity of approach. In addition, the policy instruments used for implementation vary between countries, with Italy typically including regulatory, economic and communication policies, while Croatian counties mostly rely on regulatory and voluntary measures.

Stakeholder engagement in these regions involves a variety of public and private entities, although communication often follows a top-down model, leaving potential for more inclusive decision-making processes.

The integration of these strategies into policies and their evaluation through monitoring is a common trend. All regional strategies face difficulties in finding resources and funding, while others struggle in data availability (Friuli-Venezia Giulia and Apulia Regions), employment opportunities (Marche Region) and the use of climate scenarios (Dubrovnik-Neretva County), which influence their strengths and weaknesses.

The analysis indicates that the strategies encompass a combination of "gray," "green," and "soft" measures, with the majority of funding sourced from the EU and a strong focus on long-term planning. It is noteworthy that the regional strategies exhibit a diverse array of approaches to stakeholder engagement, policy integration, and sectoral focus. The northern Adriatic regions prioritize a broad range of sectors, yet often neglect to consider cross-border risks. The central Adriatic regions demonstrate a comprehensive integration of climate scenarios and cross-border cooperation. The southern Adriatic regions place greater emphasis on mitigation, with less attention paid to transboundary risks and scientific institutions. Moreover, in all the regions, the involvement of stakeholders in the adaptation plans occurs primarily during the implementation phase rather than the design phase. This approach frequently employs a top-down methodology, with limited citizen engagement. Consequently, local communities and other key stakeholders are not adequately involved in the initial planning and decision-making processes, which could potentially compromise the efficacy and acceptance of the adaptation measures.

10.2 Synthesizing Climate Data and Strategies

A comparison between observed and projected climate impacts and regional strategies revealed some alignments and some critical gaps. The consideration of sea level rise, temperature increases, and seasonal precipitation trends is not uniform across regions. While northern and central regions incorporate these projections into their strategies, the southern Adriatic appears to have overlooked these crucial aspects. Moreover, the integration of climate indices into adaptation strategies is uneven. Although some regions prioritize the consideration of heavy precipitation days and temperature increases, there is a paucity of comprehensive planning for sea surface height and its implications.

In general, the majority of the strategies analyzed incorporate climate scenarios. However, the variables considered are often limited to a few generic examples. It can be observed that the scenarios in question do not focus specifically on the areas of interest of the respective strategies, which can result in the adaptation measures being either irrelevant or inaccurate. To ensure the efficacy of adaptation strategies, it is essential to develop more detailed and localized climate projections. This will enable the strategies to be tailored to the specific conditions and challenges of each region. This underscores the necessity for a more unified methodology for integrating robust climate projections into adaptation planning.

10.3 Implementing Strategies: An Open Database of Climate Adaptation Actions

The Climate Menu for Adriatic Regions toolkit represents a significant advancement in the field of promoting adaptive measures tailored to regional needs. The toolkit, comprising 160 technical sheets outlining best adaptation practices, offers a practical and accessible resource for public authorities and stakeholders. The actions are distributed across various sectors, with a particular focus on biodiversity, agriculture, public health, urban settlements, and water resources management. The toolkit places a strong emphasis on actions that are sustainable and that might have the synergistic effect of also being suitable for mitigating climate change. Furthermore, it proposes specific measures tailored for the studied regions, ensuring that adaptation efforts are context-specific and effective.

In summary, the various actions outlined in the Climate Menu for adapting to climate change can be grouped into three key categories that are common to all study areas: (i) community capacity building, (ii) informing and alerting the population, and (iii) organizing climate change informative meetings. These actions underscore the significance of education, awareness, and proactive community engagement in developing resilience against climate impacts. By concentrating on these foundational activities, the institutions can facilitate the development of a more informed and prepared public, capable of responding effectively to climate challenges.

10.4 Pathway Forward

The analysis presented in this book emphasizes the significance of a comprehensive, inclusive, and flexible approach to climate change adaptation.

The results indicate that key recommendations for a pathway forward include increasing stakeholder engagement through a bottom-up approach. This would ensure that local communities, businesses, NGOs, and scientific institutions are actively involved in the creation and implementation of climate adaptation strategies. This participatory process addresses the needs of those most affected by climate change, fostering ownership and commitment among stakeholders. Additionally, the integration of robust climate projections into all strategies is of utmost importance for the design of effective, forward-looking adaptation measures. The use of updated climate models and scenarios allows policymakers to anticipate future changes and tailor measures to withstand future conditions. It is also crucial to prioritize the allocation of resources and adequate funding to the most vulnerable sectors and regions, including agriculture, biodiversity, water management, and coastal infrastructure. Moreover, it is recommended that interdisciplinary planning be considered, with the objective of ensuring that all relevant dimensions are taken into account, including environmental, social, economic, and geographical factors. Furthermore, the enhancement of cross-border collaboration between Italy and Croatia, as well as among their respective regions, should reinforce the unified approach to addressing climate change, facilitate the sharing of resources, and effectively address transboundary risks. Finally, the fostering of innovation and knowledge sharing through the use of new technologies, nature-based solutions, and innovative governance models could strengthen the overall capacity to address climate change. Platforms for knowledge sharing, such as conferences, workshops, and online databases, can disseminate successful adaptation strategies and foster regional collaboration.

In conclusion, the challenges of climate change in the Adriatic coastal areas can be addressed effectively through a coordinated approach that integrates scientific insights, stakeholder engagement, and adaptive governance. By building on the strategies and tools presented in this book, the studied regions can enhance their resilience and pave the way for a sustainable future. The subsequent stage of the process should concentrate on intensifying local involvement, refining adaptation strategies, and ensuring that climate resilience becomes an integral component of regional development.

Annex

Information collected for the review of the adaptation and mitigation strategies

GENERAL INFORMATION

In this section, it is requested to provide the General Information regarding the strategy

Name of the Strategy: ____________________

1. Type of strategy:

   ☐ Mitigation
   ☐ Adaptation

2. Type of document that refers to the strategy (e.g. European Directive, national strategy, regional law …): ____________________
3. Agency that issued the strategy: ____________________
4. Year of issue: ____________________
5. Geographical area covered by the strategy (e.g. Croatia, Marche Region, …): ____________________

CONTEXT

In this section, it is requested to describe the context in which the strategy was developed. Several motivating factors explain why governments decide to develop and promote a strategy. These factors include pressures, compelling information or key events that combined persuaded governments and other influential stakeholders of the need for action. Whenever possible, it is asked to report the motivating factors behind the strategy.

6. General objectives of the strategy: ____________________
7. Description of the strategy: ____________________
8. Motivating factors of the strategy:

   ☐ Extreme weather events/impacts

E. Gioia, *Climate Change Adaptation in the Adriatic Coastal Areas*, Disaster Risk Reduction, https://doi.org/10.1007/978-981-97-6655-0

- ☐ EU policies
- ☐ Economic costs of inaction
- ☐ Scientific research
- ☐ Media
- ☐ NGO advocacy
- ☐ Private sector interests
- ☐ Recognising opportunities
- ☐ Examples from other countries/regions/…
- ☐ UN summits/reports
- ☐ Other

RELEVANCE

In this section, it is requested to point out the science and policy interactions in the development and implementation of the strategy, by examining: (a) the scientific basis around the measures, that is the connection with scientific research on climate system, vulnerability, impacts and disasters; (b) the involvement of bridging organizations between science and policy which may either fulfil a coordinating role, provide advice on adaptation policy making, or be responsible for drafting the strategy.

9. Is the strategy based on climate scenarios and projections, taking into account geographical specificities and best available science? ____________________
10. Is solid assessments of vulnerabilities/climate risks carried out to support the adaptation decision-making process? ____________________
11. Does the strategy explicitly take transboundary risks into account, when relevant? ____________________
12. Does any scientific organization/institution participate in the strategy? Which ones? ____________________
13. Do any bridging organizations/institutions between science and policy participate in the strategy? Which ones? ____________________

IMPLEMENTATION

In this section, it is requested to assess the "putting into practice" status of the strategy, whether it is already implemented or is still in progress.

14. Vulnerable sectors and topics involved in the strategy:

    - ☐ Agriculture
    - ☐ Biodiversity/ecosystem conservation
    - ☐ Coastal management
    - ☐ Communications
    - ☐ Desertification and drought
    - ☐ Emergency and rescue services
    - ☐ Electricity supply
    - ☐ Fishery

- ☐ Floods and landslides
- ☐ Foreign policy
- ☐ Human health
- ☐ Industry
- ☐ Land use
- ☐ Responsibility and insurance
- ☐ Tourism and leisure
- ☐ Transport and infrastructure
- ☐ Urban settlement
- ☐ Water resource management
- ☐ Other

15. Time span of the strategy (from the beginning to the end of the implementation):

- ☐ Short-term (up to 1 year)
- ☐ Middle-term (1–5 years)
- ☐ Long-term (over 5 years)
- ☐ Other

16. Type of envisioned measures

- ☐ "Gray" measures, based on engineering solutions (e.g. infrastructure construction)
- ☐ "Green" strategies, based on the improvement of ecosystem services (e.g. tree planting)
- ☐ "Soft" (non-structural) strategies, focused on individual and societal behavior (e.g. awareness and information, regulation and planning, management, economic and financing)
- ☐ Other

17. Description of the envisioned measures: ____________________
18. Policy instruments envisaged to implement the strategy:

- ☐ Regulatory
- ☐ Economic/financial
- ☐ Voluntary
- ☐ Communication-related
- ☐ Other

INTERACTIONS

In this section, it is requested to identify/highlight multi-level interactions in developing and implementing the strategy.

19. Which public and private stakeholders (e.g. agencies, institutions, citizens) are involved in the development and implementation of the Strategy? __________ __________
20. What are their roles and responsibilities? ____________________

21. What kind of communication flow is prevalent?

☐ Top-down approach
☐ Bottom-up approach
☐ Other

COHERENCE

In this section, it is requested to assess if the strategy is consistent with other EU, national, and regional policies, as well as with other mitigation or adaptation strategies.

22. Is the strategy integrated into new or existing sector policies? Which ones? ____________________________

EFFICIENCY

In this section, it is requested to describe timing, budget, funding and actual costs of the strategy.

23. Foreseen initial-end dates of implementation: ______________________
24. State of the implementation:

☐ Not started
☐ In progress
☐ Delayed
☐ Completed
☐ Other

25. Reasons of deviations, if any: ______________________
26. Funding: ______________________
27. Allocated budget [€]: ______________________

ACHIEVED RESULTS

In this section, it is requested to indicate the achievements of the broad objectives of the strategy.

28. Are the outcomes of the strategy evaluated?

☐ Yes
☐ No
☐ Do not know
☐ Other

29. How are the outcomes of the strategy evaluated? ______________________
30. Has been set up a body in charge of the monitoring and review of the strategy? Which one? ______________________
31. What are identified as the main strengths of the strategy? ______________________
32. What are identified as the main weaknesses of the strategy? ______________________

33. What is the main lesson learned of the strategy? ____________________

REFERENCES

In this section, it is requested to enclose links and annexes to the original documents from which the described strategy is introduced.

34. Link to the strategy: ____________________

SPRINGER NATURE

GPSR Compliance

The European Union's (EU) General Product Safety Regulation (GPSR) is a set of rules that requires consumer products to be safe and our obligations to ensure this.

If you have any concerns about our products, you can contact us on ProductSafety@springernature.com

In case Publisher is established outside the EU, the EU authorized representative is:

Springer Nature Customer Service Center GmbH
Europaplatz 3
69115 Heidelberg, Germany

The manufacturer's authorised representative in the EU is Springer Nature Customer Service Centre GmbH, Europaplatz 3, 69115 Heidelberg, Germany. If you have any concerns regarding our products, please contact ProductSafety@springernature.com

Printed and bound by CPI Group (UK) Ltd, Croydon, CR0 4YY
06/10/2026
02225878-0003